Cities in Contemporary Europe

European cities are at the centre of social, political and economic changes in Western Europe. This book proposes a new research agenda in urban sociology and politics applying primarily to European cities, in particular those that together make up the urban structure of Europe: a fabric of older cities of over 100,000 inhabitants, regional capitals and smaller state capitals. The contributors develop an analytical framework which views cities as local societies, and as collective actors and sites for modes of governance. The book examines the economics of cities, their social structures, and the modes and processes of governance. Each chapter comprises a comparison across several countries and critically examines the book's central theoretical perspective. This is not a book about the making of a Europe of cities but rather about how some cities can take advantage of their changing global and European environment.

ARNALDO BAGNASCO is Professor of Sociology at the University of Turin. He has also taught at the Universities of Florence, Naples and Sciences Po, Paris. He is the author of *Tre Italie: La problematica territoriale dello sviluppo italiano* (1977), *La construction sociale du marché* (1993), and *Small and Medium-sized Enterprises* (1995).

PATRICK LE GALÈS is CNRS Senior Research Fellow at CEVIPOF (Sciences Po/Centre National de la Recherche Scientifique) and Associate Professor of Sociology and Politics at Sciences Po, Paris. He was previously a Research Fellow at Sciences Po, Rennes, a Jean Monnet Fellow at the EUI Florence (Robert Schuman Centre) and a Visiting Professor at UCLA. His publications include *Politiques urbaines et développement local* (1993), *Les réseaux de l'action publique* (1995), and *Regions in Europe: The Paradox of Power* (1998).

Cities in Contemporary Europe

Edited by
Arnaldo Bagnasco
and
Patrick Le Galès

CAMBRIDGE UNIVERSITY PRESS

PUBLISHED BY THE PRESS SYNDICATE OF THE UNIVERSITY OF CAMBRIDGE
The Pitt Building, Trumpington Street, Cambridge CB2 1RP, United Kingdom

CAMBRIDGE UNIVERSITY PRESS
The Edinburgh Building, Cambridge CB2 2RU, UK.
http://www.cup.cam.ac.uk.
40 West 20th Street, New York, NY 10011-4211, USA   http://www.cup.org
10 Stamford Road, Oakleigh, Melbourne 3166, Australia

©Cambridge University Press 2000

This book is in copyright. Subject to statutory exception and to the provisions of relevant collective licensing agreements, no reproduction of any part may take place without the written permission of Cambridge University Press.

First published 2000

Printed in the United Kingdom at the University Press, Cambridge

Typeset in Monotype Plantin 10/12 pt. [WV]

A catalogue record for this book is available from the British Library

Library of Congress cataloguing in publication data

Cities in Contemporary Europe/edited by Arnaldo Bragnasco and Patrick Le Galès.
  p. cm.
  ISBN 0 521 66248 6. – ISBN 0 521 66488 8 (pbk.)
  1. Cities and towns – Europe. I. Bagnasco, Arnaldo. II. Le Galès, Patrick.
HT131.C514 2000
307.76′094 – dc21   99-15851 CIP

ISBN 0 521 66248 6 hardback
ISBN 0 521 66488 8 paperback

Contents

List of figures		*page* vi
List of tables		vii
List of contributors		viii
Preface		ix
	Introduction: European cities: local societies and collective actors?	
	ARNALDO BAGNASCO AND PATRICK LE GALÈS	1
1	European cities in the world economy	
	PIERRE VELTZ	33
2	Spatial images of European urbanisation	
	GIUSEPPE DEMATTEIS	48
3	Segregation, class and politics in large cities	
	EDMOND PRETECEILLE	74
4	Social structures in medium-sized cities compared	
	MARCO OBERTI	98
5	Different cities in different welfare states	
	JUHANI LEHTO	112
6	Social movements in European cities: transitions from the 1970s to the 1990s	
	MARGIT MAYER	131
7	The construction of urban services models	
	DOMINIQUE LORRAIN	153
8	Private-sector interests and urban governance	
	PATRICK LE GALÈS	178
References		198
Index		212

Figures

2.1	The metropolitan concentration in the central part of the European urban X. The squares are the European-level metropolises and the circles the national-level metropolises, according to the classifications of the PARIS Group and BfLR in table 2.1.	*page* 51
2.2	Three ideal spatial models of European urbanisation: A = balanced hierarchical (Christaller); B = unbalanced hierarchical (core-periphery); C = interconnected hierarchical and complementary.	52
2.3	Dynamic-functional classification of forty-eight major European urban agglomerations (Conti and Spriano 1990).	61
2.4	Functional connections between the main cities in the central-western Po region, calculated on the basis of distances and the cities' endowment of services (Emanuel 1988).	67
2.5	Three schematic models of the evolution of European urban centrality (in black) and its effects on regional development (in grey): A = semi-peripheral extension of the 'core'; B = hierarchical decentralisation; C = distributed metropolitan centrality.	69

Tables

		page
2.1	Some hierarchical-function types of European cities	56
5.1	Period of relative preponderance of industrial over agrarian and service employment and the year when industrial employment was at its peak in different countries (Therborn 1995: 69).	114
5.2	Income differentials in some OECD countries measured by Gini index (the greater the index, the greater inequality with regard to available income per inhabitant), in 1986–7 (Atkinson *et al.* 1995).	116
5.3	Indicators of female participation in wage labour in some OECD countries in 1986–9 (Julkunen 1992).	116
5.4	Scandinavian cities with more than 150,000 inhabitants in 1992–3 (Nord 1994).	119

Contributors

ARNALDO BAGNASCO is Professor of Sociology at the University of Turin

GIUSEPPE DEMATTEIS is Professor of Geography at the Politechnico di Torino

PATRICK LE GALÈS is Senior Research Fellow at the Centre National de la Recherche Scientifique and Associate Professor of Politics and Sociology at Sciences Po, Paris, CEVIPOF

DOMINIQUE LORRAIN is Senior Research Fellow in Sociology at the Centre National de la Recherche Scientifique, CEMS, Paris

JUHANI LEHTO is Professor of Social Policy at the University of Tampere

MARGIT MAYER is Professor of Politics and Sociology at the Free University of Berlin

MARCO OBERTI is Senior Lecturer in Sociology at OSC Sciences Po, Paris

EDMOND PRETECEILLE is Director of Research at the Centre National de la Recherche Scientifique (CSU-IRESCO) and Associate Professor of Urban Sociology at University Paris VIII

PIERRE VELTZ is Professor of Economic Sociology and Director of the Ecole National des Ponts et Chaussées, Paris, LATTS

Preface

The idea for this book was first put forward during a joint seminar held in Turin with the Istituto di Studi Superiori di Scienze Umane del Politecnico di Torino. We would like to thank, in particular, Prof. Carlo Olmo, chief of the ISSU for his support and encouragement.

The book was then prepared during one of a series of three-day seminars organised by the Observatoire du Changement social en Europe Occidentale in Poitiers in June 1996, and the European doctoral summer school that we run at the same time. We are therefore grateful to the Fondation pour l'innovation et la Prospective du Futuroscope de Poitiers. We wish to thank le président René Monory, Olivier Cazenave and Michel Roger for their continuous interest and financial support.

On the academic side, Henri Mendras and Vincent Wright provided some of the sharp, constructive criticisms that have been their hallmark for many years. We also thank the Ph.D. students who presented their work and discussed our thesis.

Nathalie Ingremeau was once again the driving force behind the project: it was a privilege to work with her. We also wish to thank Mireille Clémençon for preparing the manuscript, and the translators, Simon Lee and Neil Cooper.

The book first appeared in French as *Villes en Europe*, published in Paris by Les Editions de La Découverte. This is a slightly revised version.

We have just learned that Vincent Wright's final fight against cancer is over – terribly sad news for all those who knew this extraordinary character. We were lucky to work with him for many years and this book was the last of the Poitiers series he organised with us. We loved him dearly and his driving force, humanity and ferocious sense of humour will be sadly missed all over Europe. This book is dedicated to him.

Introduction European cities: local societies and collective actors?

Arnaldo Bagnasco and Patrick Le Galès

Europe is inconceivable without its cities. Ever since Venetian and Genoese merchants established east–west trade routes to do business with German and Flemish cloth merchants, since the bankers of Amsterdam, Frankfurt, London and Florence invested large sums in business ventures and expeditions and in loans to the nobility and sovereigns of their time, since the corporations and guilds of master craftsmen and merchants acquired or purchased charters, since boroughs were formed throughout Europe in the eleventh and twelfth centuries, cities have been the melting pot of Europe. Whether they developed within the 'interstices of feudalism' that were dominant in France and in England, or whether they were sufficiently powerful to marginalise the aristocracy, they nurtured new political ideas, capitalism in its nascent form, the bourgeoisie, the arts and culture, and individualism. Cities have shaped the imagination and the life of Europeans.

This may amount to mere nostalgia for the past: a remnant in our imaginations that is still remarkably alive but whose reality vanished with the end of the eighteenth century. After all, nation-states absorbed cities, and the building of national societies – never an easy process – was partly carried out in spite of cities by integrating them into a national whole. Furthermore, the explosion in information technology and telecommunications and the process of globalisation appear to invalidate standard individual representations of space and time. Beyond the mythical dissolution of cities under the impact of technology, the European concept of the city appears invalid. In the age of the metropolis, is the model of the European city itself obsolete?

Metropolis and cities

Urban research has reflected these changing viewpoints. For several years now, renewed interest in cities in Europe and world-wide has stressed technologies, flows, networks and the place of cities within these. This development is an essential one even if the technological determinism of some writers owes more to futurology and bears only tenuous links with the social sciences. Networks, environment, sustainable economic development, technologies and economic and financial globalisation are key phrases in leading work carried out in urban economics and geography.[1] If they are even addressed at all, political and social questions are invariably approached negatively with the emphasis on spatial and social segregation, the stages by which a socially excluded class or neighbourhood comes into being or the fragmentation and powerlessness of urban government. Problems such as these which have led to new ways of considering metropolisation have also shed light on global cities and, in considering the intensification of global capitalism, focused upon the major metropolises of Asia, America, Europe and, sometimes, Africa. The accumulation of data regarding all kinds of flows – financial, transport, telecommunications – between major cities when the argument is taken to extremes, suggests the disintegration of the notion of cities. These then become so many nodes within networks in pursuit of their own logic (O'Brien 1992). Flows and networks contribute to the concentration of huge urban regions in Shanghai, Hong Kong and Mexico City, in California or in Cairo, yet at the same time to the dispersion of activities, to the fragmenting and destructuring of cities – the very opposite of what European cities once were. The accelerating mobility of capital, goods and increasingly of people – or more accurately of certain groups of people – within the context of a growing market ethos is likely to result in the heightening of different forms of competition between cities and greater instability in urban systems.

Without questioning the reality of the effects of globalisation and metropolisation or that of the effects of technological transformation (see the concept of 'metapolis' proposed by Ascher, 1995), our view is that these questions should be submitted to investigation of a far more rigorous kind – the discourse being too general – and with more discrimination. The emphasis placed on Tokyo, London or New York tends to make one overlook the fact that in certain highly urbanised parts of the world, these 'megapoles' have only limited significance in the urban landscape.

This book proposes a rather different viewpoint and a fresh research agenda applying primarily to European cities, more particularly those that together make up the urban structure of Europe – a fabric of older cities of over 100,000 inhabitants, regional capitals or smaller state capitals as well as perhaps the huge conurbations such as London and Paris. The volume attempts a revival of questions whose inspiration is drawn from Max Weber. We endeavour to contribute to a *new comparative political economy* of European cities which gives an account of the transformations in Western society. European cities are analysed both as political and social actors and as local societies: not as metropolises, but as cities. This is how the term 'city' in the title is to be understood.

Indeed, the study of cities is frequently mistaken for that of metropolises, more especially in the United States, but also in Latin America, Asia and Africa. The theoretical current established by Park and the Chicago School has been a source of inspiration for urban sociology with its version of 'urban ecology': a research tradition which has been revived and is still of considerable importance. In view of the rapid growth and transformation of urban metropolises, urban ecology could be seen as the major problem. This approach is especially valid because of the wealth of ethnographic material. However, this material tends to envisage urban structure as merely the consequence of environmental factors, which of course corresponded to American metropolises in the 1930s,[2] yet is less crucial for the understanding of European cities.

Furthermore, the urban sociology of social interaction often hinges on the metropolis. A good many commentators have contrasted Weber's analysis of old European cities with Simmel's analysis of the *Groszstadt* at the turn of the century, modelled on Vienna and more especially Berlin. To a certain extent, the contrast between the two is less clear-cut than is supposed, and both points of view may be helpful for analysing contemporary European cities. The integrated city of the Middle Ages has unquestionably gone. Nevertheless, not all European cities have become large metropolises. Or, more precisely, the processes associated by Simmel with metropolisation have gone hand in hand with the upholding of local urban societies, and this probably accounts for the longer lasting influence of a significant trend in urban anthropology in Europe (Hannerz 1980). For Simmel the metropolis is the seat of social interaction, and his analysis of its different dimensions makes him one of the leading authors of urban sociology.[3] However, for Simmel, modern society and the social interaction it gives rise to merge with the metropolis. Social interaction is situated but the mediation of the metropolis is defined at such a level that although it embraces global society, it loses any specific territorial dimension. By stressing the microsociological analysis of social interaction and social relations within the metropolis, Simmel is not interested in social structures and intermediate policies regarding family, class or city. And, in particular, insofar as they lack the means for analysing local societies, Simmel's generalisations about cities are ultimately excessive, partial and unverified (Bagnasco and Negri 1994).

Nevertheless, it would be misleading to suggest that a systematic metropolitan bias exists in urban sociology because a good deal of classic urban sociology is concerned with neighbourhoods and immigrant communities, not to mention the vast literature on community studies.

To these two classical foundations of urban sociology one must add the critical tradition, particularly its Marxist component, which has developed significantly in all continents. This research tradition, which in the United States is sometimes known as 'urban political economy', makes it possible to reason in terms of the conflicts in social dynamics, the role of production, the processes of domination and the economic and social relations of capitalism as factors which determine the city. As a diverse tradition it remains equally fertile and influential and in some respects we are close to it (for instance the geographers D. Harvey and D. Massey).

A further major tradition in the urban sociology of the city is that of Max Weber. Weber deals with cities within his general analysis of the origins of the institutions of modern society. His view is that the city is a complete society conceived in the same way as the state, as an ideal model in sketch form, or as the Greek *polis*. Weber's analysis is fundamental since he alone proposes an analysis model for local societies and for cities as social structures, and as sites where groups and interests gather and are represented. In our opinion this dimension remains pertinent where European cities are concerned.

In this book we refer to a new set of questions affecting European cities – questions which belong to the Weberian tradition. Such a view brings back into the study of cities a type of analysis of economic and social institutions and ways of regulating interests known to the Anglo-Saxon or Italian worlds as the 'new political economy'.

Indeed, although the attention given to cities by social scientists has never faded, it is markedly gaining in intensity now for several reasons: the problems posed by the size of some metropolises; organisational and financial administration difficulties; and the concentration of new forms of poverty in deprived areas. Some contemporary social problems are today perceived as urban problems. Recent interest in cities has to do with their becoming protagonists on the economic and political stage again. On occasion, this new role for cities has produced extravagant assertions. However, there is positive evidence to show that something in their context has changed. Cities are increasingly becoming sites for the accumulation and production of wealth and they may spawn inequalities of the most glaring kind or provide a safeguard against what one knows to be the socially devastating effects of market forces. The intensification of the market produces *non-lieux*, yet paradoxically reinforces localities and regional areas, or some of them at least.

Perceiving this change has been the driving force behind this book. The assertion that cities have become political and economic actors in Europe is for the present merely allusive. The aim of this introduction is to transform this allusive assertion into an analytical standpoint, and this requires a preliminary historical overview.

Renewed interest in cities

At the beginning of the fifteenth century, Paris was the largest city in the world with a population of about 275,000. Milan and Bruges were some way behind; then came Venice and Genoa (Hohenberg and Lees 1985). These facts conceal the start of two parallel cumulative processes. In a corridor running from central Italy to the Baltic Sea, via the Rhineland and Flanders, there thrived many free merchant cities in which production and exchange were developing and capital accumulating. The traces of this Europe of cities are still clearly visible in the urban framework of the Europe of today. To the west and to the east of this corridor – in France, Spain, England and Prussia – one finds instead a gathering of military strength and the ability to exercise political control over vast territories, thereby robbing the lesser feudal lords of their power in a process that was the shaping of modern states (Tilly 1992). With the exception of Paris, which was already a capital, the major cities of the period were those of the new economy.

In 1900, the largest cities were, in order: London (6.5 million), then Paris, Berlin, Vienna and St Petersburg. Nation-states were firmly established and their major centres of power – capitals – concentrated population and wealth. The cities of early capitalism changed slowly whilst a new type of city revealed a new parallel process of concentration. The larger capitals were followed by Manchester and Birmingham; the economy had developed new technologies and new means of production. Factories and industrial cities accompanied industrial capitalism. Modern industrial society took shape across the territory.

In the course of the period in question, cities could be clearly differentiated according to their functions and their boundaries (walls and gates) from what lay beyond – villages and countryside. On the eve of the third millennium, things have become more complicated. The concentration of population has reached a level which makes administration difficult. The criteria by which a city can be defined – levels of concentration and density – are not the same everywhere. Thus a density of 1,000 per km^2 corresponds both to an urban area according to the Indian census and to the average density of rural Japan.

An urbanised area stretches unbroken beyond the administrative limits of old cities without the metropolitan zone – the true city of today – necessarily having an administrative identity. Milan, for instance, ranks both as the second city in Italy with 1.4 million inhabitants and as its first urban region – twenty-fifth in world order – with 6.7 million. The limits of a major urban region may no longer correspond to one city alone. We speak of the Rhine-Ruhr zone, comparing it to London and Paris. These transformations and uncertainties in themselves show the extent to which the organisation and functions of cities have changed.

The vast majority of the population in developed countries today live in large, medium or small cities which form the context of social life. However, with increased population movement, communication networks are assuming ever greater importance. These developments seem to lower a city's significance as a stabilised organisational context for social life; yet, in contradiction to this, cities are clearly again becoming actors (more or less as single units) both nationally and internationally. Cities create their own identities which make them recognisable abroad, specialise commercially, form alliances, reclaim a kind of foreign policy and create conditions which enable them to become nodes within a network of economic and cultural relations. Seen from this angle, there seems no lessening of the city's significance: it is changing the way it structures itself and it is up to us to change our methods of analysis.

This is why the present volume aims to examine the new urban reality of Europe comparatively, leaving aside the larger metropolises. In the context of the movement towards economic globalisation that challenges traditional forms of equilibrium, cities are now subject to significant centrifugal forces. Nevertheless – and this is our fundamental tenet – cities remain significant tiers of social and political organisation. Two sizeable fields of analysis are emerging which we shall call 'urban integration' and 'urban governance'.

The first deals with the conditions that enable cities to remain at the heart of temporally stable economic and social relations by means of actors who direct their actions mutually. Thus cities are localised societies that are structured in various ways. The second has to do with the ability internally to regulate the interplay of interests – as a contribution towards integration – and their external representation as a reconstituted whole, thus implying that cities, at least to a certain extent, constitute separate units as actors.

On the basis of this analytical pattern, our introduction states the case for a reawakened interest in cities from the stance adopted by Max Weber. More than any other classical sociologist, Weber took the city seriously as an essential element in the social structure (Weber 1921).

He considered the city as a complete local society which could be analysed through its economy, its culture and its politics, which were congruently and specifically interconnected. Regarding politics in particular, the ideal type he identified was the city capable of autonomy *vis-à-vis* the external authorities, with its own policies and constitution, and representing in a unitary way before the world at large the reconstituted interests of its citizens. These conditions enabled the city to be wholly structured as a society.

Such conditions potentially exist but rarely occur, and only, according to Weber, 'in certain historical interludes'. Political autonomy and the related ability to structure socially can indeed only fully occur when higher authorities are weak or hesitant. The purest and most complete city type identified by Weber owes its origin to the political vacuum between the crisis of feudalism and the birth of the nation-state, when the cities of the central European corridor were developing as the driving force of capitalism in its first form, competing and co-operating with one another, and looked upon with respect by kings and emperors because of the economic power they wielded. With the gradual construction of the Europe of nation-states, cities came to lose their ability to develop a structure as localised societies.

History does not repeat itself and city-states are unlikely to reappear. It is nonetheless likely that with the problems facing national states, the crisis in international relations and the construction of Europe, there could emerge a new climate of doubt and uncertainty for the higher authorities: a new historical interlude – whose stability and length we can only surmise – that may once more bring some political space to cities. This space may be limited, yet the room for manoeuvre is growing for cities. The social and political actors are aware of this and are developing their strategies accordingly.

Although the city represents a significant form of social organisation, its new prominence cannot only be attributed to the consequence of political vicissitudes. It has to be seen in the context of institutional change and changes in the organisation of contemporary capitalism. To outline this context within the scope of our argument, we shall first consider new forms of capitalism in their temporal and spatial dimensions and then proceed to clarify and develop the two fields for analysis identified above: the city as society and as urban governance.

European cities?

The existence of a significant category – 'the European city' – deserves further investigation. In *The City*, Max Weber portrays the medieval city of Western Europe with the following features: a fortress, a market, a court of justice and the ability to ordain a set of rules and laws; a structure based on associations (of guilds) and a degree of political autonomy materialised in the existence of an administrative body, along with the participation of the bourgeoisie in local government and, less frequently, the existence of an army and a genuine policy of foreign conquest. Two additional features were in his view essential: (1) rules applying to landed property (the taxes and impositions of feudalism not applying to cities) and the legal status of citizens, and (2) citizenship associated with affiliation to a guild and with relative freedom. Max Weber, as was his custom, then proceeded to give various examples that showed the singularity of the Western city. In view of this combining of political autonomy, culture, genuinely urban economy and differentiated social structure, and with its surrounding wall, the Western city was an original social structure in the Europe of the Middle Ages.

Today, any attempt to identify the common traits of the European city would seem to be a risky exercise. The influence of states has structured various urban forms. Urbanisation has been through a series of movements, and the city no longer constitutes an integrated and relatively closed system. A portion of the research being done on cities no longer bothers with such nuances and presents models that are relatively universal, or then again, the diversity within Europe appears in such a way as to rule out any kind of generalisation.

In what follows the European city will be defined essentially in contrast to the American city. A number of features are singled out: pattern and age; characteristics of the European urban system; town-planning; social structure; membership of the European Union. The picture that emerges will then be qualified and made more specific. This should be seen as a sketch to support the argument rather than a definitive model.

Morphology and age

The morphology of the European city – or its commonest version, at least – has been well described by generations of geographers and historians. Unlike American cities which are organised around a geometrical plan (the 'grid'), European cities are characterised by a built-up area around a focal point – administrative and public buildings, churches, squares and open spaces, areas for commerce and trade, and development that radiates out from this centre. Benevolo (1993: 60–3) observes four major innovations of the medieval European city which have remained: a pattern of streets and squares that bring together public and private buildings, thus creating a public space with which inhabitants can identify; the complexity (even in those days) of the various administrations which are located near one another and are easily recognisable, and the various local quarters that take shape; the enclosure of the city with walls and gates and the concentration of buildings and population within this enclosed space; and the dynamism of cities as they gradually changed. Little by little, the medieval city first evolved as a result of such development and because walls and recognised frontiers gradually disappeared to make way for *faubourgs*, i.e. suburbs beyond the walls, and peripheral urban spaces. In the built-up model of the European city, 'the density of construction is coupled with growing population density. The model is a far cry from the horizontal model of the North American city, where vacant space constantly recurs, where streets do not merely cater for the eye of a potential admirer, where open spaces frequently suggest neglect, and where – when one exists – only the business centre, which juts out as if to provoke the surrounding horizontality, asserts itself like a Mecca' (Cattan, Pumain, Rozenblat and Saint-Julien 1994: 8).[4]

The age of European cities and the relative stability of the urban system – metastability, the authors quoted immediately above call it – constitute the second classically distinctive feature of European cities. The majority of European cities came into being and were developed roughly during the first wave of urbanisation in Europe between the tenth and fourteenth centuries (Hohenberg and Lees 1985). Most of the cities of the first period of capitalism have lasted and still make up the framework of the urban system in Europe. Certainly the industrial revolution triggered a second major wave of urbanisation across Europe, but its effect was relative, sometimes only marginal, in most countries, with the exception of Britain, Germany and Belgium. The homogeneity of European cities is heightened by the fact that overall, in spite of time-lags and variations from one country to another, the major waves of urbanisation in Europe were broadly similar. Everywhere one finds medieval cities, industrial cities and capital cities.

The lasting structure of European cities went hand in hand with the remarkable longevity of their built-up form. Until early in the twentieth century, cities in Europe remained dense and organised within a relatively limited space. This longevity also means that many cities developed gradually. This in no way implies resistance to change. Wars, revolutions and crises of one sort or another certainly shaped the fate of many a city; but the system as a whole, especially the hierarchies that formed, remained remarkably stable over the centuries. This can also be seen in the preservation in a large number of cities – except nineteenth-century industrial cities or those that suffered aerial attack during the war – of historical centres and medieval or Renaissance quarters, or those dating from the eighteenth or nineteenth centuries. This longevity manifests itself in town halls, churches, palaces and other buildings belonging to one period or another according to the wealth of the city and its bourgeoisie and the influence of the state. Here urban Europe stands in striking contrast to American cities. Cattan et al. (1994) note that, although a tendency towards decentralisation and suburban development had appeared in most cities in Europe by the 1950s, largely as a result of car-ownership, and as an echo of suburbanisation in America, this observation needs qualifying: decentralisation was only really significant in the case of larger cities; its impact was no more than partial in drawing population away from the centre, whose dominance remained (true also in the cases of Paris or London); and financial and political reinvestment in historical city centres in most cases ensured their survival.

According to this first level of analysis, the classical model of the medieval European city remains alive and well. Apart from the cities of the industrial revolution, the sheer length of time over which the urban population and economy have developed has enabled a large number of cities, regardless of their various fortunes, to draw increasing advantage from the cycles of economic change. The stability of the urban system in Europe can be seen in the relative stability of their classification in order of importance, both nation-wide and within Europe as a whole. The history of the growth and development of urban Europe (Hohenberg and Lees 1985) is perfectly clear on this point. The larger medieval cities were frequently best able to absorb technological innovation, economic development and new forms of political organisation. They were able to diversify their operations and achieve growth which in spite of reverses was relatively well sustained. Excluding the effect of the industrial revolution (though here too the later phase brought considerable benefit to a number of regional capitals, e.g. Lyons and Stuttgart), main cities have tended to grow in strength during the course of history – a point largely confirmed by marked urban growth in Europe since the 1950s.

The current information and communications revolution may radically transform the classical notion of cities and their ranking; the flow dynamics are a destabilising force in this respect, and this is clearly demonstrated by Cattan et al. in *Le système des villes européennes*. In the short and medium term, the larger and medium-sized cities are best placed to benefit from current economic and technological change, and thus to buttress their positions. The hierarchy of European cities has taken centuries to come about. Radical changes may well occur (as Graham and Marvin (1996) suggest), the pace of change may well accelerate, but such changes will take a long time to carry through: 'As with any highly complex and dynamic impetus, the effect of the initial set of conditions is the deciding factor' (Cattan *et al.* 1994: 34). Where European cities are concerned, the pattern of organisation continues to have enormous significance.

Characteristics of the urban system in Europe

These structural effects are reinforced by a distributive effect. While not denying the significance of ongoing changes, one may qualify their impact in time and space. A whole area of urban research over recent years has taken shape round the notion of global cities and of the determining role of visible and invisible flows and new inequalities. Without wishing to impugn the existence of a globalisation rationale – major networks, fragmentation and competition between cities – it seems to us that a model which is a mix of 'global city', 'information city', 'entrepreneurial post-Fordist city' and 'dual city' (to quote the titles and formulations of important, recent studies by Harvey (1989), Castells and Mollenkopf (1989), Sassen (1991), and Fainstein, Gordon and Harloe (1992)) would give an inadequate account of the reality of European cities and of their transformation.[5]

Most of this research was carried out primarily with regard to the larger world cities: metropolises. But a fundamental characteristic of Europe is that there are far fewer such metropolises; they are the exception not the rule. London and Paris can probably be classed as global cities even if there are some questions about what 'global city' means. They are also the ones most subject to international competition in terms of economic development. Other European cities are conspicuously absent. Furthermore, we must realise that London and Paris are not like Los Angeles; in London and Paris the concentration of activity around the centre still has meaning, and movement away from the centre goes hand in hand with movement back towards the centre.

The larger European conurbations that come to mind such as the Ruhr, Randstad, Rome, Berlin or Milan are largely composed of cities in the customary European sense and organised around cities. Thanks to their painstaking constitution of databases on European and world cities (Moriconi-Ebrard 1993), Cattan *et al.* are able to highlight what distinguishes Europe. With a degree of urbanisation comparable to that of Japan and the United States, Europe is characterised by the very large number of cities and their marked closeness to one another: 'for an urban population that is 30 per cent higher than that of the United States, the European urbanised community alone counts three times as many urban areas of over 10,000 inhabitants (3,500 as against 1,000)' (Cattan *et al.* 1994: 23). As the major cities of Europe are not huge, one may note the small number of metropolises with a population of over two or three million, and 'if one compares the total number of urban areas of over 200,000, the average size is of the order of 800,000 in Europe as against 1.3 million in the United States and Japan ... the top thirty American cities are markedly larger than the top thirty European cities' (Cattan *et al.* 1994: 26). Europe is also characterised by the relatively high number of small and medium-sized towns. Europe distinguishes itself by the relatively large number of urban areas of between 100,000 and one or two million. In 1990 the European Union contained 225 urban areas of 200,000 or more, forty or so of these exceeding one million and a very small number two million.

This factor is of great importance in the analysis of European cities: one to be accounted for in part by their age and by the fact that they formed before the development of transport. The relatively stable core of Europe's urban system is made up of medium-sized and reasonably large cities which are fairly close to one another, and of a few metropolises, whereas the urban system in America is predominantly composed of huge metropolises which are relatively far apart. Of course, there are also small and medium-sized cities in the USA and they have a little bit more in common with European cities of similar size. The contrast here should not be overemphasised.

Political and social structures

European cities are also different from American ones by virtue of their different political and social structures. However, the ground here is trickier insofar as most European societies are firstly national societies shaped in the matrix of the nation-state. And although the nation-state now has a less central, less formative role, it continues to exercise an influence. Thus, it is virtually impossible to outline the European city today without stumbling against the distinctiveness of nation-states and the societies they fashion. Even so, certain common features can be detected if one uses the glaring contrasts afforded by the United States.

In the first place, population mobility is considerably lower in Europe. It is often reckoned that Americans move house far more often than do Europeans. Figures may vary but the extraordinary mobility of Americans is a potentially strong factor of urban instability. Conversely, the relatively low mobility among Europeans, where and when it exists, is essentially a factor of stability and continuity, favouring the constitution of social groups and public action in cities. Secondly, American society remains a society of mass immigration whereas immigration into Europe is comparatively low or moderate. Thirdly, in all European countries, central government has played a far greater role than in the United States. The impact of government on GDP is about twenty points higher in Europe (despite variations, affecting Britain in particular). This impact of the state – and, particularly, of the welfare state (unlike the residual welfare system in the United States) – has had a considerable effect on education, the reduction of inequalities and the structure of employment (see Esping-Andersen 1993; Therborn 1995). And until recently, though it has contributed to the development and stability of capitalism, government has also provided a rampart against market forces. In most continental European cities, public employment at present represents between a quarter and a good third of all jobs. If one adds to this the fact that services to consumers often count for roughly one third of employment and that the redistribution of financial resources almost never leaves urban local authorities totally dependent on market forces, there is a further structural element providing European cities with relative stability in the face of change, which in no way hinders significant albeit gradual transformations. Again, at the risk of jumping to conclusions, it is worth noting that European cities have a larger middle- and lower-middle-class element in the public sector (slightly lower in Britain and in southern Europe), and this has played a significant part in urban politics in Europe over the last twenty years.

With economic and social inequality being more marked in the United States, it was often thought that social and spatial segregation was less apparent in European cities. Put more bluntly, ghettos in the American sense of the term are rarely to be found in European cities. The revival of interest in the question of segregation and comparative studies of a more systematic kind have somewhat qualified this notion (see chapter 3 by Preteceille below and the assessment in Burtenshaw, Bateman and Ashworth (1991), chapter 3). Given that the development of spatial segregation is more noticeable in the larger metropolises and that Europe has a higher proportion of medium or medium-large cities, it follows that visibility is lower. Research into segregation has also frequently focused on the processes whereby disadvantaged sections of the population – immigrants, the poor and workers – are segregated. In this respect, social segregation is more marked in the USA – and more visible. As is made clear below, American cities are as often as not characterised by forms of inner city crisis which are only too apparent to a European observer. A comparison of the inner cities of New York, Los Angeles, London, Paris and Rome necessarily leads one to the conclusion that there are greater concentrations of poverty in the United States (Body-Gendrot 1992). The problem is less obtrusive in suburban Paris or Rome.

Nevertheless, those researching into social segregation do stress the fact that it is spreading both to the middle and upper classes. The latter in particular have the ability to select their place of residence. Here European cities (with the important exception of cities in Britain and most nineteenth-century industrial towns) differ significantly from the United States. For historical reasons, which are linked to the role of the centre in European cities, the more privileged social groups and those who constitute the political, economic and cultural elite have to a great extent continued to reside in cities and in their centres, as they always have done. True, most European cities possess exclusive residential districts on their outskirts, middle-class suburbs reminiscent of America; yet this has never precluded (except perhaps in Britain) continued residence by privileged social groups close to city centres. Comparative studies of social mobility show higher rates of upper-class segregation in European cities, and not merely in the larger metropolises (Burtenshaw *et al.* 1991).

In Europe, the bourgeoisie and aristocracy did not flee the centres of cities; they stayed, they multiplied, they accumulated capital in all forms. And, later on, those in management or the self-employed followed suit and settled either on the outskirts of town or near the centre. With the exception of the industrial cities of the nineteenth century and of ports, European cities have been less systematically affected by the inner-city crisis. And the middle classes have often been sufficiently involved to direct factory building and low-rental housing to the edges of town, although this is more the case in France and southern Europe than in Britain.

State intervention

Similarly, European cities are distinguished by the existence of public services and infrastructure and by a tradition of town planning. Both of these are closely linked to the state and to public policy. There remain as always considerable variations between one country and another as regards local authority housing, for instance (Harloe 1995). Consequently, the level of analysis needs to be defined precisely. A comparison of public services, infrastructure and planning at urban level throughout

Europe generally reveals differences between Scandinavian countries, southern Europe and Britain. Germany and France fall into different categories depending on the authors and the subjects. And one has to stretch to a greater degree of generality and abstraction, and use the contrast with major cities in the USA, before being able to identify common features.

In the first place, there is the evidence of forms of town-planning in more or less all countries: city centre renewal on the model of Haussmann, ring roads, the new cities of the 1950s and 1960s (though not in southern Europe), areas of local authority housing development, refurbishment and renovation of historic parts of the city, public transport development, the upkeep and development of parks and open spaces, industrial estates on the outskirts and, with the start of the 1970s, pedestrian and commercial precincts (Burtenshaw *et al.* 1991; Newman and Thornley 1996). Such planning reveals forms of public interventionism to counter market forces that are more infrequent in American cities. However, after the fashion of the United States, there has also been the growth of shopping malls and centres at the city's edge. A number of developments relating to the increased use of cars, while common to both sides of the Atlantic, have taken different or diluted forms.

Secondly, more than in the United States, the creation and development of networks have assumed public forms or forms that can be considered as such. Even if cities appear to lend themselves particularly to virtual interaction, their physical and material dimension is indisputable. Most European cities have seen physical investment grow over the centuries, though vestiges of the past are clearly present. Infrastructure development, and water, transport and energy networks gave the nineteenth century an obvious industrial reality. Public utilities – water; electricity; postal, transport and refuse services – were at the heart of the municipality in Europe. Indeed, one may well see them, along with the concern for welfare, as its hallmark.

These elements combine to give a fairly robust picture of the European city, so long as one focuses on what makes it specifically European – the preponderance of medium to medium-large cities. The model presented for analysis in this volume is intended to exemplify this Europeanness. It is not the Max Weber integrated medieval city, but it counts for something. As is shown here by Giuseppe Dematteis in chapter 2, the European imagination has been deeply marked by the city; the city has meaning. Our aim has been to show that this imagination was not only fuelled by the past but by a number of more or less precisely defined objective elements. Invisible flows may develop in scale, but there is still much that is positive about the majority of European cities. There are of course tensions which can call this sketch of a model into question and they are touched on below. The medium-sized city in Europe has both relevance and significance; it remains for a sociological model to be constructed.

This line of argument should be maintained, we believe, despite the fact that there are also mechanisms of concentration and dispersion in urban areas all over Europe. There are some uncertainties about the morphology of European cities in particular. We argue that despite those tensions (and Dematteis in his chapter makes it clear that he sees radical transformations in the near future) we still take cities in Europe seriously.

But now, before presenting our model for analysis, we need to describe the setting, in particular as it has been affected by the transformations of capitalism.

Forms of capitalism in time and in space

The national state was a social project within a space. It represented the political organisation of a society with its economy and its culture. In the project, politics, culture and economics had the same territorial frontiers and the same radius of action. Integration was never complete: culture circulated, some states and some economies at certain periods dominated others, but as a project the national state worked. After a lengthy process, several national states organised European society. In the words of Charles Tilly: 'It is as if Europeans had discovered that in the conditions which have dominated since 1790, a state calls for a radius of at least one hundred miles, but finds it a problem to extend its authority beyond 250.' The two figures represent the bounds within which society can be territorially organised.

Two centuries later, the difficulties which states encounter in order to maintain an acceptable territorial organisation of society have become apparent (Wright and Cassese 1996). Accelerating a process that has been underway for some time, the spheres of economy and culture have been applying pressure to appropriate new organisational spaces. The economy has been moving towards developing new spaces with market globalisation and an organisation of production which render the limits of national regulation ever more obsolete. This trend can be demonstrated by an example that shows the strength of economic pressure on the normative capacity of the state, one of its prerogatives. Today economic exchange is decreasingly regulated by national norms. Firstly, economic actors make up their own rules (in accounting, for instance) which are disseminated via transnational companies and business consultancies; secondly, these exchanges are regulated by international bodies. Frequently, states then take up these rules and incorporate them more or less as they are. In the practice of international commerce as it was formed in the Middle Ages before nation-states came into being, the *lex mercatoria* represented a legal regulation of commercial dealings which was distinct from the constitution of states or cities, and was recognised and respected by states although they had not formulated it (Galgano 1976).

Whilst the economy enlarges its sphere of activity, culture appears to restrict it. National symbols lose their significance in the face of the uncertainties and variability of economic relations. Local and regional cultures are again becoming valorising codes for the satisfaction of needs for self-expression and identity. Thus, in connection with a cultural trend, we come across an infranational level of social organisation for the first time in this analysis.

Here then is our starting point, as the renaissance of regionalism and localism has often been interpreted in cultural terms. This viewpoint is of course a simplistic one. The important question is not to know whether the rediscovery of local cultures stems from the need to express identity: this may be the case, without it having much effect on social organisation. Put in these terms, the question is reminiscent of an old problem in the sociology of modernisation, that is the resistance of traditional societies to market penetration. But economies are less and less able to consider themselves in isolation: a cultural and a political idiosyncrasy can endure only by being associated with an economy that can withstand the free market. Furthermore, there are not only forms of resistance but also, and in particular, economies capable of developing in a local matrix. Cultural resources from the past associated with a local identity can be selectively valorised in new forms of regional development: the small-firm industrial districts in different European countries are prime examples of this (Bagnasco and Sabel 1994). Where the potential of traditional cultural resources exists, economic innovations can be associated with cultural innovation as components of a local identity.

In the thirty-year period of sustained post-war growth, and in different ways depending on the country, different models of organisation and institutionalisation came into being. Their common feature was a tendency to reinforce larger systems for mass production and regulation of the economy through government intervention. Broadly, two types or directions can be distinguished: one Anglo-American, with market regulation remaining dominant; and one continental European (and

Japanese) where the market receives support from the production of public assets (like vocational training in Germany, for example), is given stability via contractual procedures (tripartite negotiations in Sweden) or supported by the development in different countries of vast welfare systems.

The American and British economies are characterised by their recourse to the financial markets, by the high level of self-financing in private enterprise, and by administrative independence *vis-à-vis* banks who have no stake in firms and are not directly represented on boards of directors. In continental Europe, on the other hand, limited self-financing goes together with the transmission of interests to other firms, cross-interests, banking interests in the capital of firms and banks frequently being represented on boards. In addition, it must be noted that the deep social rootedness of these economies in relation to ones that are more market-regulated has not come about merely by political action in the strict sense. Interest groups, associations, community relations and informal networks have all been developed with the purpose of producing stable relationships and modes of communication for the supervision and transmission of information and other resources, to develop trust and generally to make common investment strategies both possible and stable in time.

Over a considerable period, Western European models enabled increasingly rapid growth to be combined with correspondingly beneficial consequences in terms of social citizenship. The end of the postwar expansion and the first political responses have, however, made the situation more confused (Crouch and Streeck 1996). The globalisation of the economy, markets that are more unstable and increasingly differentiated and the fiscal crisis in the wake of increased public expenditure are all elements in the uncertainty that clouds our future. The response to this new situation has everywhere been to have recourse to the market.

It is no longer certain – indeed it would appear to be quite the opposite now if one compares the indicators of Germany with those of the United States – that the creation of public resources which are available to the economy, the possibility of investing in projects with a long-term return without sacrificing sectors that are not immediately profitable, and the guarantee of social harmony and cohesiveness on the basis of negotiated agreement – are deciding factors for a sound economy. The rapid movement of investment, the possibility for both firms and government to reduce costs, and deregulation that provides more room for manoeuvre, now appear as the dominating factors able to stand up to the new global competitors who are in a position to combine lower costs with quality in their products.

The return of the market renders an earlier yield on capital invested more likely and means a loss of control in the long term or less strictly defined effects in the economic circuits. The benefits of the freest market economies might be the result of the initial circumstances of large-scale structural change. For the time being, the policies of national states are coloured, in varying degrees and according to their own methods, by the new economic liberalism.

The accelerated pace of new capitalism – in reaction to the market so as to modify productive systems and shorten the return on investment – favours some forms of production to the detriment of others. It favours financial operations rather than industrial production, and the economics of non-material assets over the production of material goods. More generally, the current economic period, unlike the past, is characterised more by mobility than by organisational stability. The different institutional models also have spatial implications. The rich urban and regional tradition of the greater part of Western Europe has in the past been an important factor in shaping national models of capitalism, certainly more so than has been traditionally acknowledged. In many cases the economy was embedded in an urban and regional society.

If some cities remained by and large local societies, others lost their structure and have long been subject to national and international solutions. Conditions are again ripe for a number of cities to find a new legitimacy, a new role in economic development. The globalisation of trade and monetary flow implies that their economies are no longer embedded in a national economy. Economic globalisation signifies the increasing mobility of capital, and therefore, to a degree, the possibility of breaking free of spatial constraints. Paradoxically, this release goes along with an increased awareness of territory, of cities in particular, as potential contexts for investment and for living. This signifies a new phase in the development of capitalism, whereby capitalism itself gains an advantage over national states. The process of creative destruction involves deindustrialising cities and industrialising or reindustrialising other areas (Harvey 1989). And the competition between cities expresses the decline of state regulation and the fact that cities (in the sense of governing coalitions) are endeavouring to see where they stand, to a certain extent, in the context of such competition (Cheshire and Gordon 1995).

Ultimately, these tensions cause local society to disintegrate. Surveys show, however, that such an occurrence is not that frequent. Internal and external integration find a point of balance in spite of being constantly called into question either with a view to adjusting to economic development or to seeking protection from the damage done by the market. Conversely, the creation and rapid dissemination of innovation occur most often in cities where there is both a sophisticated research network and networks of small and large firms that are capable of providing innovative environments. The organisation of cities as actors could also be interpreted as a collective response to the threat posed by a capitalism that is oversubject to market uncertainties. More precisely, this could be a reaction to the threat posed by a deregulated market which might stimulate and foster the ability of these actors to speculate on these markets, to organise and reinforce their ability to accept some territorial pact. Cities are, above all, forms of 'flexibility insurance' (see chapter 1 by Pierre Veltz), because their environment and their diversity enable firms to cut their risks and to have access to huge resources in expertise, finance and infrastructure on the labour market which reduce their margin of uncertainty in the face of the hazards of global economic competition, itself a multiplying agent of risk. It is a difficult undertaking and one which increases intercity competition, with some that win and some that lose, and which may lead to the development of a two-speed society within the city. Perhaps it is those cities and regions which first encountered the limitations and uncertainties associated with the loss of autonomy that best typify contemporary societies. Two examples may serve to illustrate the part played by cities in the new mechanisms of institutional regulation.

Developing market deregulation works against traditional industrial policies. Today, however, industrial policy may centre on an area or a milieu and strive to make different investments in a locality coherent. The fiscal crisis is everywhere causing a cut in welfare benefits. A transition towards regional norms and standards accompanied by a regionalisation of taxation, together with the targeted and selective management of the missions placed in the hands of local welfare systems, seem to represent the way things are moving.

Economic globalisation is likely to produce ever more urban concentration in Europe but without any immediate disruption in terms of ranking even if relations between cities may change (Cattan *et al.* 1994; Cattan 1993; Veltz 1996). Surveys conducted in Europe largely confirm this trend, especially for the major European cities (Parkinson *et al.* 1992; Cattan *et al.* 1994). In France, Britain, Portugal and certain regions of Scandinavia (but not in Italy), there appears to be a discrepancy between a region in relative or total economic decline and the relative prosperity of a city – a regional capital, for instance – where development continues. Thus one analytical trend describes global or world cities as being those that benefit from the trend towards the concentration of command points in the world economy, especially head offices of banks and transnationals as well as the production sites of various services (Sassen 1991; Knox and Taylor 1995).

Then, if economic globalisation involves the development of trade, flows and networks, so the infrastructures that link the separate units are crucial. But the economic rationale of networks (whether they be for transport or information) is to maximise their application between significant units so as to absorb and render the costs of heavy technology-intensive investment profitable. Cities take on an essential 'conglomerate' role for infrastructure-linked industries (environmental, for instance). Here again a rationale which is outside the scope of the national state induces further concentration round major conurbations. Finally, as regards the labour market, the tendency for couples to both be in work and the ranking of labour markets from one city to another produce a concentration of the more highly waged (particularly in the private sector) in cities.

Today, the increase in the mobility of actors also signifies the marked spatial mobility of investment and forms of organisation whereby networks are established across different countries and continents and can be rapidly restructured as needs and opportunities arise. For local economies all this has long been a source of tension, and it is even more so now. After all it is open for this or that actor to find an associate in a different place, in a joint venture for instance, thus severing all links with local economic actors. And over a long period this will have an adverse effect on public assets and on social capital which is the outcome of complex, reliable and smooth relations at the local level.

Clearly cities are too frail as structures to sustain the setting up of new institutional models in isolation. The examples available show degrees of compromise between different territorial levels of government.

Cities as localised societies, social interaction and structures

Cities may be more or less structured in their economic and cultural exchanges and the different actors may be related to each other in the same local context with long-term strategies, investing their resources in a co-ordinated way and adding to the social capital riches. In this case the society appears as well structured and visible, and one can detect forms of (relative) integration. If not, the city reveals itself as less structured and as such no longer a significant subject for study: somewhere where decisions are made externally by separate actors.

Broadly, with a view to developing the analytical model, two main lines can be distinguished in the social structuring of contemporary societies. The social category of people depends upon their professional situation. It may also provide a basis for political aggregation. But one must also realise that political intervention aimed at economic regulation and various forms of economic incentives has complicated matters. If individuals are considered as producers of goods and services, the pertinence of social groups is more evident. But if they are considered as consumers, the dividing lines between social groups may be very different.

Studies carried out from this second angle during the period of continuous expansion brought about the formation of a large middle class composed of the middle classes, workers and office workers who were more alike as a result of their increased purchasing power, with more uniformity and stability as well as an approximation of lifestyle and consumer tastes centred on leisure and free-time activities. Social welfare systems, in the broad sense, which have always chiefly benefited the middle strata – rights and benefits acquired in connection with housing or home-buying and the relevant legislation, the accumulation of assets and cultural advantages, opportunities for saving and investment – brought them social citizenship. Conversely, those who did not gain access at the right time, bearing in mind what has changed over the last twenty years, may well find themselves among the excluded, hence the risk of ending up among the groups of the socially deprived in one or another European city (those who are sometimes seen as forming an underclass). Without overstating the case, this second dividing line might be seen in terms of status stratification, a term traditionally relating to particular ways of life, and socially recognised and sometimes politically safeguarded consumer potential (indeed with regard to the Middle Ages this is the sense attributed to urban status with rights of citizenship as opposed to the feudal world of the countryside).

Given this dual distinction – (active) producers/consumers, class/status group – we may try to imagine a social structure and highlight the relative developments in time. Following this line of argument, the effect of industrialisation and mass consumption has been to simplify the class structure with one vast category of blue- and white-collar workers lumped together in large factories, offices and shops. This development has made the division by class in industrial societies the major factor of social regulation and articulation of political demands. It has taken different forms in different countries depending on institutionalised traditions and the concrete nature of the economy. The enhancement and alignment of wage terms, the development of savings and capital investment – in houses, stocks and shares, etc. – linked to the growth of systems of welfare protection and, more generally, benefits that are politically administered, have further served to increase the importance of lifestyle and consumer habits – of status – as the basis for articulating political representation, more especially in certain countries and with a view to specific problems.

With the end of the period of sustained growth, individuals have again become more exposed to market contingencies. Their living conditions are more clearly and more directly related to their labour-market and class situation, but not in the same way as in the past. The new forms of business organisation and the development of the service sector have made for greater differentiation between social classes, which means that the structure of social groups has become more complex and more difficult to construct and reconstruct both for sociologists and social actors. We can no longer be satisfied with a generic definition of social classes. We need to call upon all the knowledge available to us on the economic sociology of Western societies, including the issue of social class-making and remaking.

This is where cities as local societies come in. In our view, the structuring of class has to be reconsidered by sociologists (although some have not forgotten it). Moreover, its relevance was more evident in times when cities were distinguishable by their function as capitals, and as trading, market or industrial centres. As yet we are unfamiliar with the new functional types, and therefore analysis is difficult. The city, on the other hand, has been the place for observing the processes of consumption and social reproduction, and that has drawn attention to the status group. Urban sociologists have studied political processes and collective social movements in the areas of housing, education, health and amenities, and have found that individuals take action not so much as workers, employees or management but as householders or tenants, taxpayers or parents of schoolchildren (Saunders 1986). Studies of this type are complemented by work on urban poverty and social movements.

Nonetheless, the structuring that results from the interaction of status and social groups remains an open question whilst our understanding of social movements might improve analytically, considering the political economy of cities. The position is complicated by the fact that mobility is higher than in the past, both daily and at weekends, whether for work or leisure; and indeed city users who move around for services or as consumers have grown in number. More and more individuals 'use' or frequent cities without having links that would give them responsibilities, such as being taxpayers, etc. (Martinotti 1993). This is very much the case in larger cities and is an element in the disintegration of the fabric of urban society. In a more general way, the largest cities are probably too complex and differentiated to lend themselves to structured analysis. On the other hand, the large city is particularly apt for studying the dynamics of social interaction. As Simmel pointed out (1903), pressure for the development of individualism is particularly marked. It is an old question. Is Simmel's highly individualised person, with his problems, a creature of the metropolis or does he in the first place belong to this or that stratum of society? Is he a product of industrial cities or of major capital cities? Is he simply a product of Berlin? In concrete analysis, interactive models need to be combined with models of local societies. Research into segregation and poverty, as into ethnic groups and districts in major contemporary metropolises, may be redirected with a view to more explicit structural analysis. From this viewpoint, the characteristics of urban governance may also be considered as revealing social structure.

In the medium-sized cities of Europe, social structuring is more obvious and analysis easier. Of course, there are very different city types, but in general the middle strata and middle classes acquire particular prominence and a marked capacity to shape the economy, culture and local politics, and this provides an opportunity to emphasise the importance of structural analysis at a local level for a fuller understanding of the social structure of a country (OCS 1987). For instance, if the middle strata have similar quantitative weight in two countries, but if in one the urban structure is dominated by large cities whereas in the other medium-sized cities have greater importance, then the middle strata will play a far larger economic, political and social role in the second country.

Seen from this angle, regional or urban identity ultimately consists in an implicit or explicit venture, developed by particular local actors who consider it to their advantage or convenience to pursue mutual action. This rules out neither conflict nor competition but it does imply a relatively stable context in which to interact, with schemes generated by several actors, investment in the local society with deferred returns and the creation of common assets and collective goods. The concepts of 'civicness' (Putnam 1993) or of 'social capital' (Coleman 1990) aim to account for the active cultural component in this type of case. An identity defined in this way relates, however, to a strategy and to an economic and political structure centred in a local society. And so the possibilities open to local society have come back to the question of the organisational and institutional models of contemporary capitalism.

Urban governance

Urban elites endeavour to make the city into a collective actor, a social and political actor possessing autonomy and strategies. The process involves attempts to reinforce or to create a city's collective identity, a willingness to promote a local society, especially as national identity is becoming blurred. Nobody is really taken in by the exercise; everyone knows that Western societies are not conducive to the emergence of cities as local societies in the medieval sense. Yet this does not exclude either a stronger mobilisation of interests, groups and institutions to develop a collective urban strategy to counter other cities, the state, Europe or market forces, or an attempt at redefining a kind of local culture. Certainly, individuals are subject to influences that are too complex for one to be able to believe in a system of local values that could be passed on. Even so, if government plays a less central role and if we find national societies being called into question, cities may become one of the places where local subcultures and/or social groups grow stronger in particular configurations. However, in some cases, in the long run, political choices systematically favour social cohesion or, on the contrary, play into the hands of the major institutions and firms. Political or cultural traditions are subject to reinvention, or are mobilised around a collective local project.

But even if there is no determinism at work there the territory, and in particular cities considered as a political and social construct, may constitute one of the intermediate levels at which actors, groups and institutions are structured. It seems to us that in the restructuring process that is taking place between government, market and civil society, which is discernible in the way that frontiers have become blurred (a process intensified by European integration), the extension of the market rationale, also within the public sphere, has in some areas and in different political forms produced the demand for a level of political and social organisation other than national. The spread of the market appears paradoxically to be leading to a return of the political infrastate territories such as cities. However, the reinforcement of social and political organisation in certain cities, which the concept of governance is an attempt to explain, is different from the political as defined in terms of legal and rational domination. It is much more a case of the mobilisation of social groups, institutions and public and private actors forming alliances and collective projects with the twofold aim of attempting to adapt to economic change and of counterbalancing somewhat – or even protecting themselves from – the effects of the market. This assumption is in line with the arguments put forward by some commentators who choose to stress the transition towards post-Fordism (Amin 1994; Swyngedouw 1992), though the conclusions we draw are not the same.

Put another way, the advance of globalisation and of European integration and the calling into question of nation-states and their ability to run the economy and society are giving rise to two types of development in cities at the same time: (1) as shown by geographers, cities and regions are losing their significance, and the ever multiplying flows and networks that run through them make any kind of social coherence or territorial politics illusory; (2) emancipation from the state has led to the mobilising of localised political and social energies, thus giving rise to the notion that certain cities or regions can become political and social actors and give renewed coherence to an area of territory in promoting public action, and even – though here probably in only a limited number of cases – in boosting economic development.

Consequently, one can imagine modes of governance in which fragmentation is dominant and where forms of regulation count for little. The aim of urban sociology that is tackling the question of governance is to show how these different forms of regulation operate in European cities and to show the interplay of social groups and movements, interests and urban servicing firms in the constructing (or not) of a collective actor.

Governance is thus defined as a process of co-ordinating actors, social groups and institutions in order to reach objectives which have been collectively discussed and defined in fragmented, even nebulous environments. Following the example of what has been argued concerning the governance of the economy (Campbell, Hollingsworth and Lindberg 1991), several types of regulation (or governance mechanism) can be identified. By shifting the question of the governance of the economy to the local or regional area, the aim is to be rid of the constraints and questions raised by institutional economists or the sociology of organisations, and so go beyond problems of effectiveness and co-ordination, in order to reincorporate a specifically social and political dimension. The study of urban governance provides an understanding of how the mechanism of economic and social actors who are outside cities is in part constrained and at the very least influenced. Acceptance of the term refers to what takes place beyond an organisation, namely the ability to organise collective action and to build coalitions and partnerships directed to specific ends. In this event, the problem of efficiency is distanced and it becomes necessary to introduce the different types of legitimacy, the power struggles and the creation of identity. It is not a matter of knowing what is the right form of governance but of identifying the mechanisms and processes which enable (or not, as the case may be) a more or less significant, more or less structured form of governance to be obtained. In this context, urban political systems, public–private partnerships or corporatism are merely particular forms of agreement, particular forms of urban governance.

So, by way of example, Bagnasco and Trigilia have coined the term 'neolocalism' to characterise the local system as it is observed to function in the Third Italy,[6] i.e. 'a particular division of labour between the market, the social structures and, increasingly, the political structures, a division which allows a high degree of flexibility in the economy and rapid adjustments to market variations, but also a redistribution of social costs and real benefits from development within the local society' (Bagnasco and Trigilia 1993). That corresponds more or less exactly with what we understand by 'governance'. It is quite clear why these two authors settle on the notion of 'neolocalism regulation'. As sociologists, they have watched these districts emerge without any intervention on the part of government, and in governance there is similarly the notion of piloting and of steering (Jessop 1994). From local regulation one moves gradually on to questions of governance in places where the role of local authorities, political institutions and political elites has not been fundamental in organising economic development. The political factor existed in their analyses but was of secondary importance. However, this factor plays – or seems to play – a larger role today (Le Galès 1998). This sociological perspective on forms of local regulation carries on – though differently – from the (unfinished) attempts in British (Cooke, 1988; Harloe, Pickvance and Urry 1988) and French research into localised social change to make manifest a 'locality effect'.[7]

One of the key questions concerns local actors, their resources, their strategies and the way in which they interact. A growing amount of European research is now giving prominence to the growth of systems and coalitions associating different types of actor. The problem as it is defined everywhere is how these actors are organised, to what extent they are rooted in cities, their degree of autonomy and legitimacy, and their strategy. The range of actors in question has grown considerably. Traditionally, comparative studies considered local authorities, thus enabling emphasis to be given to the distinction between Scandinavian municipalities built around the organisation of social and public services, the southern European model with its low political capability, and the cases of France and Britain. Virtually everywhere the political potential of urban mayors within national political systems is growing stronger. The mayors of Paris and Lisbon have been elected president, there are many mayors who are government ministers. More fundamentally, the effect of electing mayors by universal suffrage, in Germany and Italy for instance, has been to reinforce their legitimacy. Nevertheless the factors which weaken the political capability of governments clearly operate in the case of cities too. Mayors everywhere endeavour to build coalitions so as to encourage economic development. Consequently, systematic studies are starting to be made on Chambers of Commerce, employers' associations, associations of other kinds, unions, property developers, urban servicing firms (Lorrain and Stoker 1995), universities, agencies linked variously with government and various other institutions. Certain social groups are more or less organised in cities and command a degree of activity which makes it desirable for elected representatives to be brought into collective projects. Such social groups have a vital role in some cities. The city as collective actor tends to emerge out of the interplay between these actors.

This view has been taken up sometimes a little hastily in order to reveal the effects of competition between cities. The new post-Fordist urban governance should naturally lead to a combined mobilisation merely for the purpose of economic competition and to bring about a reaction, for instance, in the form of new social movements (see Margit Mayer's chapter 6 below).

The European situation is far more diversified and complex (Harding 1996). Some Scandinavian cities – Helsinki, for example – are still little, if at all, committed to this logic of competition and have limited strategy for economic development. The preservation of social services, the struggle to limit social segregation and the maintenance of social cohesion continue to dominate urban policy. The city remains structured by government and social services, and to a lesser extent by private actors. The same resolve to shield the local society is apparent in Italian and, in some cases, French and German cities. Conversely, some cities are dominated by an alien system of political or market regulation and they are neither actors nor localised societies. Urban governance remains weak and fragmented as in Paris or London, where economic competition is dominant. In other cities urban governance is evident as well as a still relative integration of diverse groups and actors organised with a view to adopting an overall strategy; Barcelona, Bologna and Rennes may well be cases in point. At the same time other cities – in Britain or, in some instances, France and Germany – tend to follow the American model of strong recourse to economic development and limited concern for the underprivileged.

The above examples point to various types of differentiated urban governance in Europe, which we must give an account of and recognise in terms of possible changes. They nevertheless indicate that cities are gaining greater political importance, particularly in terms of their strategic and political capabilities, as well as a certain kind of political approach towards negotiating with other cities, regions, governments, firms and Europe.

Presentation of the book

Sociological and political science research on European cities is still, in most cases, carried out in the context of the nation, or else it deals with the larger metropolises or represents case studies where there is little co-ordination. The point of view we adopt is at once different and more restricted, with one exception: that it leads us in many respects to propose a new angle of research and a 'new comparative political economy' for European cities, combining economic change, social structures and questions of governance.

As sociologists we could be related, on the one hand, to what R. Swedberg (1987), in his history of economic sociology, has called 'new political economy', which takes politics and social structures more seriously and examines the ways in which the state and conflicts between interests, classes and institutions tend to regulate the market and organise societies. On the political side, the 'new political economy' approach takes on board the role of the state to support capitalism but also focuses on the role of the state to regulate or govern the market, to protect society from the destructive effects of the market, and to develop institutions and public policies in order to constitute a protection for groups and territories within the borders of the nation-state. This sort of approach has been particularly developed in order to study the regulation of the economy. Since Weber and Polanyi, there has been interest in the political and social foundations of the economy and in the regulation mechanisms that differ from the market mechanisms to explain the transformation of Western societies and of economic development. For instance, Hirschman (1981), after Marx or Polanyi, has highlighted the destructive effects of the market on societies but also the conditions under which the market may reinforce or even create societies. Authors from that 'new political economy' stress the autonomy of the political dimension, the importance of the power struggles between groups and institutions, and their role in the bid to regulate the economy. Taking an interest in the issues of regulation and governance amounts, a priori, to an examination of the regulation of social and political order.[8] That new political economy approach is therefore reluctant to see the city as passive space. As mentioned before, urban elites endeavour to establish the city as one collective actor – a social and political actor endowed with autonomy and with strategies. This process involves the attempt to reinforce or create a city's collective identity and consciously promote a local society, and that leads to various struggles between groups as some hegemonic projects are contested.

However, as urban and economic sociologists, we have developed in this introduction a neo-Weberian analysis of cities as local societies. It is the combination of the two – cities as local societies and cities as collective actors and sites for modes of governance – which is at the heart of our intellectual project. We are fully aware of some limitations in our approach, for instance the risk of reification associated with the presentation of cities as collective actors. Cities rarely act as such, of course, and it is necessary to keep that in mind and to analyse the strategies of individual and collective actors. We have tried to argue that it is worth taking the risk to develop a new comparative political economy of cities in Europe.

The book is an attempt to test part of our hypothesis with a group of European colleagues. This is a first and partial attempt and there are some discussions in the book about the validity of our theoretical orientation and the methodological consequences. The book is therefore not a complete and co-ordinated research project to support our approach and we have not tried to cover all European countries. This is not the point. Several avenues are explored in the book and all chapters are written from a comparative point of view in order to critically examine some parts of our framework. Not all papers cover all European cities: very far from it as small-scale comparisons are often the way forward in our field. We are fully aware that some parts of Europe are not well represented and that some chapters only cover two or three countries, not to mention the absence of Eastern European cities which are so fascinating.

We have followed our Weberian inspiration with a division of the book into economy, social structures and issues of governance.

The first section of the volume is concerned with the major trends in urbanisation and urban economics. In the first chapter Pierre Veltz considers 'archipelago economics' and the role of the city in the face of the changes in productive organisation. If European cities offer advantages in slowing the economic process, Veltz concludes, which in itself may give meaning to urban governance, they are also geared to the more pronounced forms of globalisation. Giuseppe Dematteis, in chapter 2, adroitly reassembles the images of European cities and what is made of these images, while setting out the main direction of globalisation. He is willing to concede that medium-sized cities in Europe still count for something (perhaps for ten or so more years, he suggests), but in his view the real challenge consists of directing thought and enquiry on to metropolises of every kind, especially those in a network system.

The second section is given over to cities as social structures. In chapter 3, Edmond Preteceille contrasts the arguments for spatial segregation in the larger metropolises. At the same time he tackles questions of social structures, in particular highlighting the contradiction between the amassing of wealth and the social and political fragmentation of major cities. In chapter 4, Marco Oberti comes to grips with the problem of social structures in medium-sized cities in Europe and, on the basis of this, outlines a typology of cities. Juhani Lehto, in chapter 5, examines the ways in which welfare states in Scandinavia contribute to the social and political structuring of cities.

The third section of the volume deals with problems of governance. Margit Mayer (chapter 6) examines the development of social movements as urban actors in European cities and their contentious role in the processes of urban governance. In chapter 7, Dominique Lorrain takes examples from the history of urban services to challenge the notion of the impotence and disintegration of the political factor faced with large-scale technological systems and firms. Urban government has room for manoeuvre when such systems face crisis and certainly when there is a need to define roles for corporate action. In chapter 8, Patrick Le Galès tackles the same question with reference to private economic actors. Their larger role in the processes of urban governance may signal either a defeat for local politics and increased fragmentation or else the emergence of collective action in the processes of construction.

European cities are again in the centre of the stage. The intellectual aim of this volume and of our research is to break with a view of the city considered only as metropolis, without in any way denying the importance of such a view. All European cities have a 'metropolis' dimension to them in the way Simmel understood it, in that they provide a density of interaction, favourable conditions for a monetary economy, and autonomy for the mind and for the individual. To a certain point they are also still partially integrated urban societies. Disintegration and fragmentation in cities tend to occur in varying degrees and according to context. Our view is that in the context of Europe, cities still have meaning. If we can justify the usefulness of such a view and the relevance of the fresh approaches shown in the following chapters, our aim will have been reached.

Notes

1 See, for example, the series of contributions in J. Brotchie, M. Batty, E. Blakely, P. Hall and P. Newton (1995).

2 On this point see Bagnasco and Negri (1994). Martinotti (1993) has proposed a new analysis from the point of view of urban ecology.
3 Simmel (1965, original edition, 1903) and for a discussion of his work see J. Rémy (1995).
4 For a shrewd commentary on these points, see R. Sennett's fine book (1992, French edition).
5 By way of example, the application of Sassen's model to Paris and London has provoked an interesting debate. C. Hamnett (1995), E. Preteceille (1996) and P. Veltz (1996) have variously produced economic or social arguments or shown the role of government and the welfare state in shaping cities – arguments which function differently for the United States and for Europe.
6 The expression 'Third Italy' refers to the north-east of Italy (in particular Veneto) and the central regions (Tuscany and Emilia Romagna), i.e. not the *mezzogiorno* or the industrial north-west. Over the past thirty years these regions have been characterised by remarkable economic growth organised within industrial districts in small and medium-sized forms – a mix of market mobilisation and communities.
7 See OCS (1982, 1987).
8 See, for instance, in that tradition R. Boyer and R. Hollingsworth (1998).

1   European cities in the world economy

Pierre Veltz

Cities in the economy, between the rise of the global and the return of the local

In the early days of political economy, from Richard Cantillon to Adam Smith, the city, the interaction of cities and the relation between city and country were active categories of economic analysis. Then, with the exception of the occasional unorthodox approach – that of Jane Jacobs, for instance – the categories became somehow passive and were covered by the homogenising film of the 'national' or 'international' economy. However, the predominance of national scales for economic performance – backed up by the fact that today national perimeters are far and away the best served by statistics – calls for reflection. The increasing interdependence of so-called 'national' economies is in fact accompanied by persistent, frequently increasing, regional – i.e. infranational – disparities. The most significant cases of development are not so much national as limited to particular, sometimes restricted, zones which are invariably under the control of large cities (Chinese growth provides an excellent example).

In Europe, as elsewhere in the world, one can see the accelerated concentration of production and consumption in metropolitan urban areas, both within the orbit of an outsize city – the Paris or London regions – and of a more dispersed kind, as in northern Italy and the Rhineland. The extreme polarisation of 'global' financial activity and of high-tech research within a few world centres, the dynamism of the new city-states relieved of the costly problems of a hinterland – Singapore or Dubai – are merely extreme forms of a process which at different degrees concerns all developed economies. One increasingly has the impression of an 'archipelago economy' in which horizontal, frequently transnational, relations increasingly outmatch traditional vertical relations with the hinterland. What explanation is there for such polarisation at a time when the growing fluidity of communication and the volatile – in particular, uncapitalistic – character of the most remarkable activities in modern capitalism ought to be leading towards a more and more footloose economy, in which places and concentrations are a matter of indifference?

This first question immediately prompts a further, more political, one: is there not a connection to be made between the rise of urban economies and the increasing difficulties encountered by nation-states to define and implement coherent and efficacious economic policies and, more generally, maintain their role as first-rank social and cultural reference points? Following the unambiguity of a centuries-old triumph of 'territorial' over 'urban' economy (to take up the terms used by Braudel following K. Bucher), while the techniques of supervising and homogenising space, themselves decisive in the building of modern nation-states, have become both commonplace and obsolete, there is a temptation to see the present situation in terms of revenge on the part of cities over states. And indeed, the re-emphasis of locality in a political as in a cultural sense has been a significant sociological factor for a number of years now. But history moves forwards not backwards. If reference to the city-states of the Middle Ages and the Renaissance contains a mythic potential, it has little to do with present economic realities. The 'urban economic policies' of the pre-modern era were fundamentally protectionist (Weber 1982: 163). States have long leaned upon urban economies by taxing them, while taking care not to destroy them. But the equipoise showed signs of instability with the rise of big-firm capitalism. Cities found themselves irredeemably engulfed in an open national and international economy where they counted more as complex nodes in networks than as isolatable entities. So the problem now is to discover how localised politico-cultural energy can harness support from economic forces that reach far beyond the local sphere, indeed rather dominate and direct it. We shall see that there are two possible responses: one purely reactive, involving politics and culture as a means towards recapturing an identity that economics has tended to dissolve; the other more subtle – the urban actor not merely showing assertiveness and/or winning over nomad investors, but actualising the worth of historical and territorial structures as competitive resources in far-reaching economic networks.

So to a third question: in this interplay of localised and global is there anything specific about the European city? It has to be said at the outset that we hardly possess the empirical data required to deal with so difficult a question. As regards external form, it is known that the texture of cities in Europe differs from the cities in North America, if only because there are more small or medium-sized cities and the contrasts in urban authority are slighter (though more marked in France and Britain). Marked inequality of urban density sets England, the Netherlands, the

Rhineland and Italy against the rest of Europe (Pumain, Rozenblat and Moriconi-Ebrard 1996).[1] Attention can further be drawn to the social and symbolic place of centres and certainly the existence of characteristic forms of city living that are immediately discernible, if difficult to put one's finger on. In other respects, calling to mind the original political shaping of the European city – the 'conspiratorial' association of bourgeois against the 'legitimate' powers, in Weber's outline[2] – has little meaning now. What common ground is there between the megalopolis of the present, whether dispersed or concentrated, and the tiny Italian city-states of the Middle Ages? Hence, a further hypothesis, which needs of course to be put to the test, is here presented. What perhaps in the economic sphere characterises the European city is a particular capacity to mobilise – within the market economy itself – resources that are generally considered to be outside the economic sphere – shared cultures, social networks of multiple types, structures of cooperation – in line with methods that soften the impact of the more brutal and impersonal effects of the unadulterated market. Such intertwining of economic, social and territorial factors was probably a major influence on growth during the long post-war expansion in France – even if, in economic analysis, the role of cities and regional areas was almost entirely overshadowed by insistence on the macroprocesses of the welfare state.[3] In present economic conditions, however, there is nothing to show that the undivided extension of market regulation is the only competitive course. The accelerated globalisation of the economy finds expression in complex and ambivalent processes. On the one hand, it unquestionably does constantly offer new fields for market deployment – deployment whereby the specificity of territory is often disregarded or destroyed. On the other, are we shall see, the profound transformation in modes of competition – where the so-called 'non-costable' factors of competitiveness, such as the quality of goods and services, occupy a decisive place, particularly in countries with a strong currency and high welfare protection – potentially reinforces the role of non-market interaction, social institutions, and forms of co-operation, trust and experience that have been accumulated and so to speak stockpiled across the territory. More than ever, these socio-historical elements constitute the 'hidden face' of market competitiveness; and this serves at the same time partially to restore strategic space to local actors in economic as well as in political terms.[4]

The 'archipelago economy': some trends

The concentration in metropolitan areas of jobs, particularly the most highly skilled jobs, and of growth is a global feature to which Europe is no exception. Despite all the methodological reservations one may voice about spatialised computations of 'productivity', the survey undertaken in France by Laurent Davezies is remarkable in its findings. Not only is productivity per individual at work considerably higher in large cities than in those with a population of between 100,000 and 200,000 (25 per cent higher in Paris) but it grows faster (+30 per cent between 1982 and 1991 as against +12 per cent) (Davezies 1996). This has to be seen in relation to highly selective forms of mobility. On average the French have shown little mobility over the last twenty-five years; however, mobility among young active people and that of top-level jobs together produce figures that are highly favourable to Paris (Julien 1994). Southeast England shows similar results.

A second trend reflects the relative despecialisation of regional and urban economies. The profiles for regions and cities sector by sector are tending slowly but surely to converge. Territorial differences are now seen more in terms of levels of employment than of specific sectors. Clearly this is reinforced by a climbing tertiary sector, which as a random category tends to homogenise the profiles of regional sectors. But every analysis confirms that 'services' are becoming increasingly segmented by level of employment, less and less by type of activity. The process of European unification may well find partial expression in the reinforced specialisation of certain 'poles' (in the Netherlands, for instance, with maritime commerce and logistics), but growth is particularly focused on areas endowed with a high degree of internal diversity and that proffer a range of activities that are relatively close.[5]

A third trend is marked by some disconnection between centre and periphery, together with a rise in horizontal relations between major 'poles'. In this respect France presents an interesting case. Between the 1950s and the 1970s, the major development involved massive decentralisation of medium- or low-skill jobs in industry (in both large-scale industry and small and medium-sized firms which became increasingly integrated into the subcontracting system) to the benefit of regions where in the past agriculture or cottage-type industry had prevailed, at the same time as posts at a creative or supervisory level remained centralised in the Paris region. Thus an unequal entity came into being with which the centres and peripheries were closely interdependent. It is precisely this unequal coupling which is now giving every appearance of slackening. Factories in the provinces face continuous job losses and upwardly moving activities – communications, etc. – are becoming concentrated in the major cities or in areas previously unindustrialised, e.g. the Mediterranean coast. The most active regional 'poles' rely far less on their periphery than on direct relations with other 'poles', with Paris in particular (in the case of France, direct relations between major regional cities are weak). The case of Toulouse in this respect is not untypical.

The growth in horizontal relations is also seen at the international level. The growth in air traffic between major focal points such as Paris, London, Frankfurt and Amsterdam is much faster than elsewhere (Cattan 1993). Incidentally it needs to be emphasised that none of this implies an automatic decline in rural or in interstitial areas. Paradoxically, one of the indices of obsolescence in the traditional model of the centre *vis-à-vis* the periphery is that the contrast between rural and urban worlds is decreasingly relevant, not just in highly urbanised countries such as Germany but also in countries with low population density such as France. The profiles of employment and population in rural and urban areas are now very close. Cities no longer drain the countryside of manpower, some rural areas prosper while others stagnate, frequently without there being any apparent geographical reason to explain it (Mendras 1988, chapter VII; Hervieu 1993).

To this picture should be added the essential but unrecognised role of localised economic activities linked to public funding. In France, and very probably elsewhere in Europe, there is a geography of market activities and of public-sector or semi-public-sector employment, with the latter dominant in a number of cities (particularly in the south and depressed areas of the north), and there again the focusing of public resources bears little relation to the traditional centre–periphery pattern.

A fourth trend is discernible in the growth of inequalities between zones, between cities and regions, and indeed within the same urban areas. For historical reasons, Europe presents regional contrasts that are far more marked than in the United States.

But such inequalities (average wages, income per household) were greatly reduced following the second world war for the very reason of patterns of mobility – of labour but especially of capital, which moved preferentially to where labour was cheaper.[6] All indicators now show that the spread of these disparities is remaining constant, if not widening. Even within metropolitan areas, however true it may be that 'dualisation' is less significant, and in the theoretical case of the 'global city' put by Sassen (1991),[7] surveys still show an appreciable and general increase in inequalities between localities (Davezies 1995).

These trends need to be complemented by the growing space–time differentiation affecting individuals. Even if residential mobility remains fairly limited, intermittent mobility has developed apace for a large part of the population, whereas many continue to be narrowly confined and their horizons remain hopelessly limited.

All in all, the overall economic and social picture would seem to be increasingly difficult to chart with the use of simple spatial diagrams, on the familiar model of pyramid-shaped representations of structure that stack together like Russian dolls; yet it is this model we continue to favour in social and, even more, in political contexts. We must reconcile ourselves to this. The world is no longer well ordered by distance, clearly 'layered' (to borrow Braudel's image) between short- and long-span economies. Even so, the generalised presence of the global economy with its rhythms and constraints in no sense ushers in a world in which territory is in some sense neutralised, if not cancelled out. But the territory that counts is more and more the territory of social interaction, not merely of physical proximity.

Globalisation, competition and change among productive organisations

Some cities evidently are further advanced in the process of economic globalisation than others. But, whereas the thesis of 'global cities'[8] is focused exclusively on a few world metropolises and a few leading-edge sectors (finance, top services linked to transnationals), the significant point is that economic globalisation now affects the whole range of cities and territorial structures.[9] It affects them not merely through sectoral variations (downturns here, expansion there, direct competition in the location of this or that product), but in particular through changes in production and trading methods, a consequence of the growing interdependence of economic space and changing competition.[10] These general changes and the way they have affected industrial organisation and networks merit brief attention here. It is a necessary parenthesis before considering the relevance of such changes to territory.

First, what are we talking about when we talk of 'globalisation'? The economy has in fact for a long time been global; and the break in the direction taken by internationalisation, if it has occurred at all, belongs more to the middle 1980s than now. Whilst the level of trade has grown continuously since the end of the last war, what is significant has been on the one hand deregulation and financial globalisation, itself far ahead of industrial globalisation, affecting the real economy, and, on the other hand, the spectacular rise in direct transnational investment which since 1985 has progressed much faster than international commerce or global production (insofar as one is able to measure it). Thus between 1986 and 1990 the rate of growth of such investment was 28 per cent per year, and that of commerce only about 12 per cent.[11] Europe occupies a central place in these flows (even with the omission of internal trading within the EU, itself very vigorous), being both the principal source and the principal recipient of direct foreign investment. It is worth commenting here that this mechanism of cross-investment is in the main restricted to developed countries and dramatically sidelines the poorest countries on the planet. Even though the domestic share of the economies (domestic production and consumption) is still largely preponderant, a transnational mesh of production is being formed which now makes the perception of a territory made up of self-cohesive zones trading with others something of an anachronism.[12]

This multiform, multidirectional opening up of 'national' economies has wrought a huge change in terms of competition, in particular for the larger firms that belong to the era of mass production. Overall these had a reasonably quiet existence in the post-war period, protected by frontiers but cushioned too by transport costs, national distribution networks and unchanging consumer patterns. However, new oligopolistic competition which started up in the 1970s and got into its stride in the 1980s and 1990s at an international and frequently global level marks a profound change, being both far more intense and far more unstable. It would scarcely be an exaggeration to say that a superior stage of capitalism is now making its appearance in the competitive and 'market' world, whereas such capitalism (especially in Europe) developed very largely away from competitive conditions, invariably bypassed by the dominant actors and secured for the minor ones, as Braudel has well shown.

The evident result of this transformation is that the major firms – just like the minor ones – now have to practise competition in terms both of cost and of differentiation. Quality, variety, capacity to respond, services, plus innovativeness, are now mandatory and this makes for a more complex competitive equation (see, for instance, the question of 'quality' which has threatened the very existence of the European motor industry). Naturally this applies all the more to strong currency areas (Japan since 1975, Germany and France at present). Put simply, the process may be summarised by remarking that competition was traditionally 'imperfect' because of geographical barriers, *de facto* or *de jure*; and that their removal leaves firms with no alternative but to substitute geographical product differentiation (Jayet, Puig and Thisse 1996).

The important point here is that to survive in these conditions requires a major organisational and operational overhaul on the part of firms, and particularly large firms, in their dealings with the web of subcontractors. Traditional organisations built upon rigid functional division of tasks, at the same time allowing a clear-cut and stable spatial division of labour, have become counter-productive. The common element in the imperatives of differentiation – quality, variety, responsiveness and innovation – is that they all require far greater cooperation between the stages of product conception, manufacture and commercialisation, between the earlier and the later phases, between the 'direct' and 'indirect' actors in the productive process, and between the firm and its environment. In other words, competitiveness results more and more directly from 'relational' impact, which is not easy to programme or quantify, and less and less from the traditional one of 'productivity', obtained through the intensification of tasks or activities considered in isolation. This has opened the way for a whole range of experimentation, going far beyond the adoption (particularly in the motor industry) of the Japanese principle of 'lean' production, which in itself might be regarded as the final manifestation of traditional Fordist organisation.

Three main tendencies in organisation appear to me to be evolving: the search for forms of organisation linking actors in productive chains transversally, either around the physical flow (logistics, just-in-time and so on), or through informational systems, or within 'project'-type organisations of a more or less ephemeral, variable geometry type; the combination of reinforced strategic centralisation and a degree of operational decentralisation, more especially in the context of a 'cellular model' linking more or less autonomous multifunctioning units (network–firm) in a horizontal network; the relinquishing of *a priori* normative regulations in activity and an emphatic return to regulation by objective and by results, of a pseudo-market or market type.

It will be apparent that all three tendencies contain potent sociological implications. They also share the characteristic of distancing traditional professional distinctions which situated wage-earners both within the industry and on the labour market, and gave them a clearly defined social identity. A further common feature is that of combining an explicit appeal to wage-earners' motivation and co-operation with increased stress on contractual-type regulating (assessment by results, temporary and impersonal nature of ties). This, of course, goes along with the wave of externalisation, increasingly pronounced recourse to temporary jobs, and the rise in self-employment. All over Europe, the 'market' (in various forms) is regaining ground in all areas to the detriment of traditional graded and/or co-operative forms of regulating industrial systems. This is even more the case in the services sector, and particularly in fast-developing sectors such as communications and entertainment. Naturally, exposure to all this is experienced differently by different individuals, some clinging on to professional structures which are more or less under threat while others have clearly decided to go with the way things are moving and exploit the call for general skills on a short-term basis – the more effectively because they are thus distanced in relation to collective referents and professional interdependence.[13]

In fact it would seem – though the hypothesis would require further testing – that all these trends, however unequally represented by country or region or sector, have the effect of converging in the sense that they cut across conventional socio-economic divisions. For example, there is evidence of a growing overlap between the manufacturing and 'tertiary' sectors, not merely from the viewpoint of objective economic relations but also from that of organisational methods and probably of lifestyles and work patterns too. Similarly, there is convergence between the world of large firms and that of small firms. Of course, national differences are still appreciable. In France there is not the same separation as there is in Italy between areas of small and medium-sized industry and those of large-scale industry. Economic and spatial interconnection is close, whilst the social and cultural distance between large firms' senior management and the heads of small firms is still great. But beyond these differences, which should not be played down, there are shared tendencies at work. The notion of a radical contrast existing between a 'mass production' pattern and a 'flexible specialisation' pattern has over the last twenty years or so been refuted in all countries and in all sectors. Rather, the developing context has shown gradual convergence between the two patterns: between the model of the firm as network and that of a network of firms.

From economic change to spatial forms

How is the link to be established between these economic changes and spatial representation? What one must try to convey here is the sense less of a body of linear one-to-one causal relationships than of a complex field of tensions, trends and counter-trends.

Some economists lay stress on the more or less direct impact of the differentiation of goods and services on urban concentration. In my view, this is secondary. People do not turn to a large city so as to have access to a more open, more diversified, consumer market, since this is becoming more accessible everywhere. On the other hand, the combined impact of changes in the organisation of production and in the labour markets is a decisive one. I shall summarise this in four points.

With the expansion of spatial competition and the transnationalising of production and markets, the economy clearly distances itself from territorially based localised societies. The explicit competitiveness within networks over sites for production, and sometimes even sites for product conception, understandably attracts attention, all the more so when the growth of large firms and their networks is mainly effected through external growth and when the response to market globalisation generally consists in an organisational globalisation which calls into question traditional multidomestic systems (by which each region is provided with functions that are virtually the same and juxtaposed and so superfluous). Strategic and directional centralisation but also the increasing centralisation of purchasing, i.e. relations with suppliers, are key factors in these changes. This frequently leads to a spectacular explosion in the geography of suppliers, on a European and maybe on a global scale, an explosion which is only in part counter-balanced by technological needs from the neighbourhood, as in certain forms of 'just-in-time'.[14] The extraterritorial distancing of the economy is indeed real and sometimes brutal. Yet it does not lead to a pure economy of flows that is indifferent to territorial considerations. There are powerful counter-tendencies.

The growing role of human capability – as against technology crystallised in the form of machines – is the first one to mention. It needs to be recalled that, of all the factors of production, labour is the least mobile, especially in Europe (where geographical mobility is much lower than in North America).[15] Overall, there has been nothing less than an inversion of the indices of mobility for productive resources over the last ten years. Capital which was once compartmentalised has become hypermobile (although the link between savings and investment is for the most part internal in the major world regions). Merchandise circulates more and more easily. Technology, on the other hand, which in earlier phases of mass production may well have appeared to be easily accessible and transferable (insofar as this was understood to mean relatively simple machine technology) has become markedly territory-based, because the essential technological elements are now incorporated not in machines but in human capability. Furthermore, it needs to be stressed that the function of human capability in the area of competition has to do not only with 'non-industrial' sectors but with manufacturing sectors and more generally with any activity involving highly integrated and specialised automated systems. Thus, in industry, it is the sound use of physical capital in its high-tech form that is now decisive in calculating production costs. Such use is of course a direct factor of the level of competence of the operators concerned and of their back-up (maintenance, R&D, etc.), which explains, one might add, why production of goods is often better – and cheaper – in high-waged countries.

This is all the more true when it comes to product creation and innovation.

To this it should be added that the abilities which 'deliver' in an economic sense seem to be characterised by two apparently contradictory features. They are closely related to a specific social and cultural background (in other words, they have a marked collective and relational dimension); at the same time they are 'freer' in the sense that they tend to be wildcard, detached from any particular technical system or narrow professional category. And this of course facilitates the kind of recourse already mentioned to externalisation and the networking of production systems, and favours the vast and highly differentiated labour markets of the larger urban areas. In direct line with the preceding point, it is the 'relational' nature of the modern economy which gives it its most significant territory-bound factor. The term is not used in a vague or metaphorical sense but – as mentioned above – to signify that all components in competitiveness at the present time share the characteristic of exploiting the complexity and quality of co-operativeness within the productive chain – inside the firm, between firms and customers or users, and between firms and institutions – in contrast to the segmenting and compartmentalising that typified time-and-motion study in large-scale industry. In other words, competitiveness is now increasingly 'systemic' or 'environmental' in character (the latter term being taken in a very broad sense). This at bottom explains why the interconnection between the market economy and a whole variety of social non-market forms, which have historical and geographical origins, are holding their own and indeed playing an increasing role in economic success. Trust between actors in chains of production for instance is still an essential factor in efficiency, even when the market is dominant. It is a powerful force in boosting collective responsiveness and leads to economies in organisational costs, insurance costs and so on. However, trust cannot be taken for granted in an increasingly open world. In this respect, locality as a cross-contact network is a precious resource. Certainly territory is not the only base for co-operative possibilities. All forms of diaspora, like religious or ethical communities, are shown to be equally effective.[16] Furthermore, one has to draw a distinction between two poles at least in the relational economy: that involving routine-established 'technical' relations and that of more complex interacting, which gives rise to a fuller sharing of intelligence and visions for the future. Hence a fairly widespread scheme for the spatial division of labour might be the following: production might take place almost 'anywhere', given impeccable logistics and a skilled workforce; product-conception, on the other hand, is recognised as needing specific environmental conditions and labour 'market'.[17] So, above all, it is better to avoid making a distinction between 'market' (or 'market'-type) relations and extra-market relations, since they seem to be increasingly entangled.

A final major dimension in the relation between patterns of production and territory has to do with time-scales. Increasing uncertainty – in the full sense of the future being both unforeseeable and less and less amenable to projection – the acceleration of tempo and the rise of short-termism mark a development that is transforming not just industry but the whole of our society. This development is very directly linked to financial globalisation especially in the Anglo-American model, itself still distinct from what Albert (1991) terms the 'Rhineland' model but continuing to gain ground. The spatial implications are complex and several:[18] they may be summarised as having two main thrusts. The first is the search by all means available for short-term flexibility so as to enable the worst of the impact of economic fluctuations to be evaded. Externalising and networking point in this direction. And here large cities facilitate a permanent, flexible, rearranging of chains of production, in that they play an essential role of substitution. With individuals, this implies the erosion – now endured, now accepted – of projects affecting both career and lifestyle which are by definition long lasting. Where loss of security is experienced negatively, it also implies new strategies with the object in the short term of making the most of every advantage in the labour market, at the same time as registering the little value now attaching to seniority in industry (de Coninck 1995). The second thrust is the built-in reversibility of options in the medium term, i.e. the priority accorded by firms to choices – more particularly, in regard to locality – without long-term commitment and themselves open-ended. The larger reservoirs of employment in densely populated areas have here an insurance function, there being a greater probability of firms eventually finding the workforce they need, in skills as well as in number; also there is less of a problem if it comes to disengaging. Corresponding attitudes are found with individuals, with the added effect of qualifications – which are the modern, mainly urban form, of employment insurance, and which replace the web of contacts and recommendations by word of mouth, etc. – and of two-waged households. Female employment rates are directly correlatable to urban size, and the disproportion between the possibilities offered by metropolitan labour markets and the rest is considerable, especially where highly qualified women are concerned.[19]

It should be emphasised that such recourse to the short term and rejection of the future harbours serious tensions for individuals whose need for slow progress and security are denied, and also for the economy itself, whose requirements of continuity and of memory frequently go unrecognised. Further there is the added risk of conflict between wage-earners who are temporarily protected and those who endure the uncertainties which firms will not or cannot take responsibility for.

Conclusion

Finally, it is important to stress the ambivalence and the complexity of the relations between the economy and the city which we have touched upon above. It is possible, of course, to make a broad contrast between the 'relational' (or 'environmental') model and a 'predatory' model, in which firms, in particular the major firms, simply appropriate passive – human and material – resources within a given area (the predatory model being shown to good effect in the policies of large-scale industry during the long post-war boom). But to use a clear-cut contrast between market mechanisms and non-market interdependence so as to categorise the relational model would be simplistic. The very age of their industrial traditions, their low population mobility and their institutional wealth, and the significance of their accumulated public amenities and resources, are very probably particular assets for European cities and regions in the non-market underpinning of their economy. Nevertheless, this does not mean that the larger European cities of today are spatial economic communities, industrial 'districts' given a metropolitan dimension. The close linkage between the metropolis and the rise of the most impersonal features of bureaucracy or the market, of 'objective culture' and monetary ties – a linkage to which Simmel at the turn of the century gave vivid expression – is certainly deeply ingrained in our society. Likewise, on the question of time-scale, though it may be true that being rooted in historical and territorial development is of great advantage in slowing down the pace, and in safeguarding memory and continuity, it is no less true that cities are also, more than ever, formidable machines for accelerating every type of flow and for merging identities and interaction.

Salais and Storper have produced an analytical table which helpfully deconstructs over-rigid distinction between the 'market' and 'non-market' spheres by proposing a typology of 'worlds of production' based upon implicit understandings between the actors (relating, in particular, to types of product and future anticipation) (Salais and Storper 1993). In their view, such worlds might also characterise spatial economic identities of different scales. For my own part, I am struck by the overlapping and intermixing of these worlds, especially within large cities.

The homogenising effect of mutual understandings is always subject to corrosion from other, more ephemeral, types of relation. So it would seem that one could arrive at a useful analysis of types of economic co-ordination and co-operation through distinguishing three major poles.[20] The first relates to the *contract* – whether market or not – which is impersonal or nearly so, limited in time and sanctioned by an outcome. This type of relation, far from becoming extinct, seems to be very much on the increase. Yet contracts are not drawn up on a blank page, and participants refer to *rules* and *conventions* which represent acquired knowledge and more generally shared cultures. This model has been fully studied and illustrated in the literature on industrial districts. But as a model of rules shared it does not take into account dynamic development in which highly contrasting cultures are merged and where actions are not evaluated in relation to previously defined norms, unless it be that the action itself is successful. Such development resides rather in the mobilisation of *networks* that are diverse, emergent and forward-looking, whereas contractual relations value uniformity and hence are backward-looking. The modern city, in my opinion, derives its dynamism from its ability to interweave these different methods, achieving a mix of the cold self-interest of the contract, the reassuring warmth of shared cultures and the often cynical imaginativeness of networks. A further set of remarks by way of conclusion concerns the strategic possibilities afforded by urban communities and the decisive importance of the institutions of urban governance in development. The high mean level of infrastructural amenities – transport, telecommunications, education – in Western European countries extends the field of choice when it comes to firms locating. Infrastructural considerations no longer weigh in this respect unless they happen not to be met. But correspondingly new margins for manoeuvre are opening up for the economic development strategies of cities and regions, since they are far less constrained by the facts of geography. Therefore, the least material factors of development come to the fore; and they serve to explain why some cities develop rapidly whilst others decline, in spite of their evident geographical advantages. Clearly, elements which are not at all easy to measure, such as prevailing atmosphere or local capability for project-forming or setting up a coherent collective agenda, now play a decisive role. Given the fact that crucial resources are now put together by the community rather than given as such, it follows that the quality of urban governance – whatever the precise meaning given to the term – is doubtless the foremost factor in development. But clearly this hypothesis remains to be scientifically tested.

Notes

1 When you link towns with a population of more than 10,000 that are within 25 kilometres of each other, there is a striking contrast in densities. See also Le Bras (1996).
2 Le Bras (1996), chapter 2. One should bear in mind that, for Weber, the study of cities is not a study of urban growth, but one element in a comparative study of the birth of the bourgeoisie in Europe and of modes of domination (Bruhns 1995: 107–21).
3 With the exception of analyses such as those which began to emerge at the end of the 1970s on the 'Third Italy', the 'urban districts', centres of innovation, etc.
4 This set of problems is developed in Veltz (1996).
5 These trends seem valid for the whole of Europe. See, for example, the national monographs in Rodwin and Sazanami (1991).
6 With marked national variants: for instance, the contrast between the motor industry in France and in Italy.
7 On this point see the special number (22–23) of *Sociétés contemporaines*, 'Ségrégations urbaines', L'Harmattan, Paris, 1995, and, in particular, the contributions by Hamnett and Preteceille.
8 See Sassen (1991) but also Friedman and Wolff (1982) and King (1990). The customary term 'global city' sometimes serves to mask fuller analysis.
9 Taking the long historical view, the significant feature is probably that the difference between the more 'global' cities and the remainder is diminishing rather than increasing.
10 For a fuller account of the thesis, see Veltz (1996).
11 Between 1990 and 1993 the rate fell slightly, then picked up again.
12 Thus it is estimated that roughly one third of the flow of international trade is within transnationals, a further third concerning an establishment that is part of a transnational.
13 The point is rightly emphasised by Bagnasco (1996) in regard to wage-earners in 'non-industrial' sectors, but to a certain extent it applies also to other sectors.
14 For instance, in the motor industry, attention is frequently drawn to the spatial proximity of certain 'synchronous just-in-time' suppliers. But the converse predominates, with the median distance between suppliers and assembly plants increasing.
15 Low manpower mobility is also noticeable at an international and global level. Whereas merchandise flows (in proportion to GIP) rose by 15 per cent (among developed countries) between 1970 and 1990 and capital flows doubled, population movement fell by nearly one third.
16 Weber's analysis of the part played by Protestantism in the development of the American economy remains in this respect entirely valid.
17 The pattern in the pharmaceutical industry, for example.
18 For a fuller treatment, see Veltz (1996), chapter 9.
19 In 1990, there were 350,000 women in higher professions in the Ile-de-France (two thirds of them in the private sector) as against 160,000 for the combined total of cities with a population below 100,000.
20 Here I am indebted to the work of F. Eymard-Duvernay.

2 Spatial images of European urbanisation

Giuseppe Dematteis

How can we question the spatial images of urbanisation?

There are two ways of approaching the spatial images of the city which geographers and cartographers produce and which the media, urban marketing experts and politicians use increasingly as a surrogate for analytically founded discourse.

The first way consists in accepting the evidence and the presumed referentiality of the images which, representing precisely localised individual objects, appear to us to be true as a whole, i.e. even for the not strictly geographical relations and meanings that they imply and suggest. For example, the 'European megalopolis' does not only say that between London and Milan lies the greatest European urban concentration, but also that this forms a system, which generates force fields and gradients, so that the destiny of cities depends on how they are placed compared to it: this is a partial truth, which hides many fundamental aspects of the problem, with the result that it legitimises distorting territorial and urban policies, such as those that make everything depend on major transport infrastructures.

The second way is conscious of the fact that spatial images do not reflect reality as it is, but are a mental construct, a means to highlight certain facts or certain relations that are more or less consciously linked to certain intentions, or at least a certain general vision of the problems. This second approach moves in two directions. One, which we could call *deconstructive*, aims to investigate not what is represented, but the social relations (in the broad sense) of which they are (or could become) the matter, means or conditions. The other is, instead, essentially heuristic: geographical images, thematic maps and geo-referenced statistics are a means to explore new elements that rise to the surface. They can suggest spatial relations or orders that do not come under given conceptual definitions, neither are they envisaged by consolidated models or theories. For example, the multiplication of long-distance connections between cities in the recent phase of 'globalisation' has suggested, as we shall see, thinking of the cities as nodes of networks with variable geometry in a discontinuous virtual space, in addition to being segments of territory in a continuous physical space, governed by relations of proximity.

By deconstructing consolidated images, we go back to the unspoken premises of common thought; by exploring alternative meanings, we propose new ones. Thus, the deconstruction of given images and the exploration of new spatial orders are only apparently diverging directions. In reality, both converge towards the production of feasible projects: projects, in other terms, that do not ignore the force of the images and ideas that tradition has given us, but without believing too strongly in their solidity, so as to give back to representations of reality the fluidity that enables the difficult passage from the present to the future.

Two factors of viscosity should therefore be borne in mind: a general one which depends on the fact that geo-cartographic images tend by their very nature to emphasise permanent features; the other is the specific nature of the city, which is endowed with exceptional territorial stability. It is sufficient to note that most of today's European cities already existed in the Middle Ages and many have occupied the same position since antiquity. Today this inertia tends to be seen as an obstacle to the innovative transformations demanded by global economic and geo-political competition. It seems that the cities remain the only relatively fixed points in today's hypermobility of flows and actors. This is true as physical places and must be to some extent as institutional and political entities, if they want to play a role as collective actors. From this point of view, the real stability of the city, just like the fictitious one of spatial representations, can also be a positive value and a resource for urban governance.

This essay will explore from the points of view mentioned above the vast production of spatial images that characterise the recent studies of European urbanisation. To this end, we shall start from a long-term historical vision and then move on to the more recent transformations and typologies that have been proposed. Reflection then follows on the ways of representing current urban transition as the interaction between local and global spatial levels, and their capacity to express the complexity of standardisation, interconnection and diversification processes that exist between European cities and within each of them. The cities will thus be seen as nodes in a network which, considered on the continental scale, will suggest some interpretations of the structure and dynamics of, and the prospects for, European urbanisation. In relation to this, the problem of territorial forms and units will be discussed, which is pertinent today for urban governance.

Images of European urbanisation in the long term

In 1822, the geographer Karl Ritter wrote: 'Africa appears as a limbless trunk, Asia has ramifications on three sides, while Europe seems divided in all directions, with the limbs prevailing over the body.' This interpenetration of land and sea is the geographical metaphor that best reveals the identity of our continent. It can be equally true for its cities, because they too are scattered 'in all directions' and, at least originally, 'with the limbs prevailing over the body'. The picture that emerges is that of an X, as proposed in a recent European study (De Roo 1994). The image withstands statistical assessment: still today, more than half of the cities of European importance are deployed on this X, with particular concentration in the segment that runs from the Italian peninsula to the British Isles (see figure 2.1).

Historically, this recalls the Mediterranean origins of European urbanisation and the decisive role of the Rhine valley, together with the regions on the North Sea and the Baltic, in the great medieval blossoming of the cities, linked above all to intercontinental trade and network interactions between the cities (Hohenberg and Lees 1985). But since the end of the Middle Ages, this centrifugal, networked and articulated figure has been overlapped by a centripetal one, more stocky and compact, which has preferred the trunk of the continent. It reflects the principal role that European cities have assumed in the modern era: that of go-betweens between the great national spaces in formation and the new systems of political, social and economic control. To perform this function, the cities were structured in a hierarchical system of 'central places' that tended to balance out; within each state, these reached up to six or seven levels. This was recognised and formalised in the model of the *Zentralen Orten* by W. Christaller (see figure 2.2A), when this spatially balanced configuration had already been overlapped by another, born out of the industrial revolution, which generated great imbalance: that of negative centre–periphery gradients (see figure 2.2B). This latter model, which still dominates the continental model (figure 2.1) shows on the various scales – from the local to the regional, national and continental – the characteristic of spatial concentration, which opposes the cities to the countryside, the great urban poles to their regional peripheries, the dense European core region to the less urbanised peripheries.

In the last twenty years, however, there have been changes, starting from precisely these central and semi-central areas of the continent. They are variously recorded in the more recent terminology of urban spatial transformation. The figures of disurbanisation, counter-

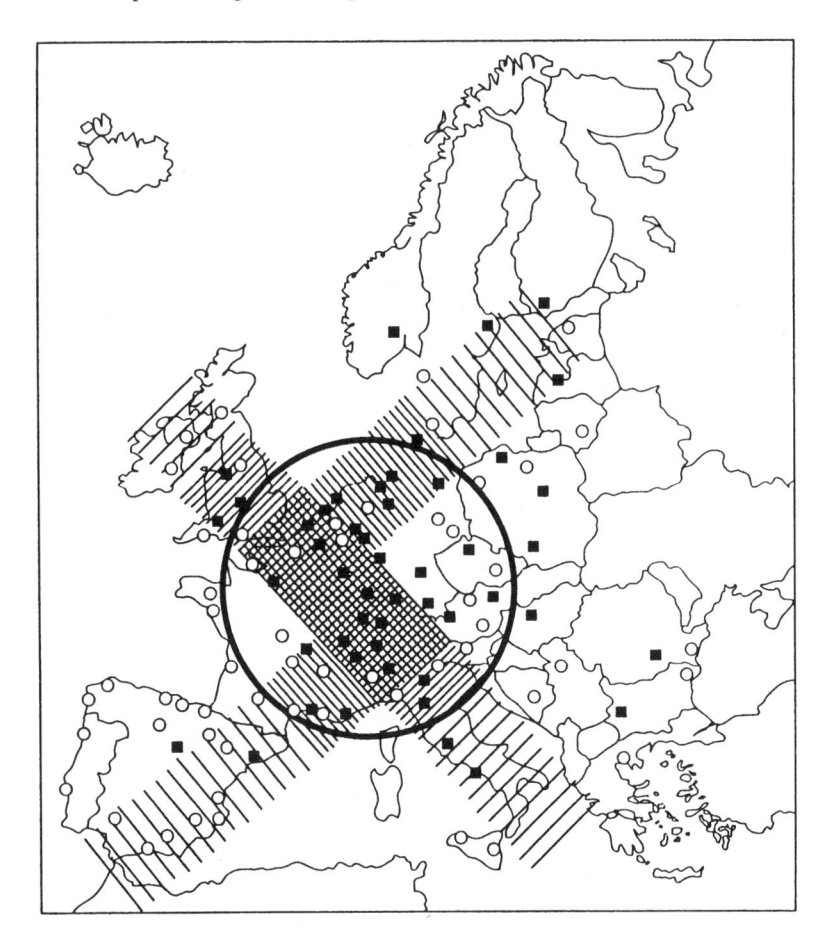

Figure 2.1 The metropolitan concentration in the central part of the European urban X. The squares are the European-level metropolises and the circles the national-level metropolises, according to the classifications of the PARIS Group and BfLR in table 2.1.

urbanisation, peri-urbanisation and others that will be mentioned converge towards the image of a new network development, which supersedes the Christallerian territorial hierarchies and the traditional core–periphery hierarchical relations, in that it rejects the idea that the organisation of the territory is based solely on relations of spatial proximity. The new networks (see figure 2.2C) multiply within themselves the horizontal and vertical relations far beyond the old bonds of proximity, bringing out the idea (difficult to depict) of a virtual space of flows no

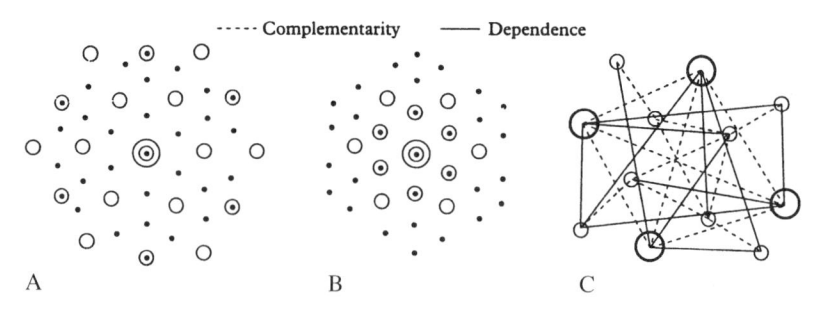

Figure 2.2 Three ideal spatial models of European urbanisation: A = balanced hierarchical (Christaller); B = unbalanced hierarchical (core–periphery); C = interconnected hierarchical and complementary.

longer conditioned by physical distance: a topologically discontinuous space, made of 'nodes' (cities) and stable connections between them, even if 'immaterial' (technology transfer, financial transactions, business and institutional agreements, etc.).

Will these new configurations allow a return, even if in different forms, to the medieval network, spatially articulated with multiple urban centres also distributed on the periphery of the continent?

Recent images: counter-urbanisation and the city life cycle

Up until the 1960s, 'urbanisation' was a synonym for urban concentration, of migration from the countryside to the cities, from small towns to the big cities. For more than a century, Europe led this trend. In 1800, of the world's hundred largest cities, sixty-four were in Asia and twenty-nine in Europe. In 1950, Europe had thirty-six against thirty-three in Asia and eighteen in North America. But in 1990, the percentages of Europe (twenty) and North America (thirteen) had fallen in comparison to the rapid growth of the southern continents and Asia was again at the top with forty-four cases (United Nations Centre for Human Settlements 1996).

The reversal in the trend between 1960 and 1980 involved all countries in the industrialised world and has created a crisis in the spatial images of urbanisation. This was described by the American, B. Berry (1976), as 'counter-urbanisation': 'a process of decentralisation of the urban population that implies the passage from a state of greater concentration to one of less concentration'. It would thus seem to be (as the rather emphatic name indicates) an epoch-making turning point, which forces the separation of the concept of urbanisation from the image of spatial concentration. A. Fielding (1982), more analytically, studying the phenomenon in Europe, defined it as the inverse correlation between the demographic dimension of the cities and their migratory growth rate. And this, at least until the mid 1980s, constituted an almost general rule (Champion 1989). The troubling image of a European population fated to be concentrated progressively in a few gigantic metropolitan areas, against which town planners had struggled in vain in the first half of the century, was replaced by another, more comforting one, in line with the contemporary ideology that 'small is beautiful', which sees in the opposing process of urban deconcentration the beginning of a natural territorial equilibrium.

It should be noted that Europe is the continent in which the conditions for the geographical redistribution of the urban population is the most favourable, having inherited from the past a very high density of cities. Excluding Russia, the average distance between urban centres with more than 10,000 inhabitants is 16 km against 29 km in Asia and 48 km in the United States. This is also because small and medium-sized cities (with less than 200,000 inhabitants) are very numerous: 60 per cent of the urban population lives in them, against 45 per cent in the United States (Moriconi-Ebrard 1993).

There were others who saw the demographic decline of the great urban centres as a phase in a medium- to long-term swing, in which concentration and deconcentration alternate. In particular, the model of the 'city life cycle' (Van den Berg *et al*. 1982, taken up in Cheshire *et al*. 1989) considers three concentric rings for each functional urban region (FUR): that of the core, the suburban ring or hinterland, and the rest of the functional region. This hypothesises that, over time, demographic growth shifts from peripheral areas to the centre (centralisation), then from the centre towards the periphery (decentralisation), then returns to the centre again (recentralisation).

The passage from centralisation to decentralisation has occurred throughout the European Community from the 1950s to the 1960s, even if not at the same time, first in the northern countries and later in the southern ones. Between the 1980s and the 1990s, the heralded recentralisation did not happen, however, at least not in the form expected of a generalised recovery of the core of the major cities and a corresponding weakening of the minor centres in the surrounding region (Cheshire 1995). There was, instead, a hybridisation of the two opposing trends, in the form of peri-urbanisation or 'deconcentrated centralisation'.

Before examining these new figures it is worth pausing to consider the meaning of counter-urbanisation and the city life cycle. The thesis upheld here is that these spatial images functioned as 'seismographs' – sensors that detected underlying movements in the economy and society – during the 1970s. When they were used as simple descriptive models to explore the new emerging socio-spatial models, they contributed to the empirical understanding of the processes. When, instead, they attempted to function as 'spatial theories', they did not help to understand change, limiting themselves to enclosing the description of open and irreversible processes within the old scheme of neo-classic economics, which reduces any dynamic to swings around a presumed natural state of equilibrium.

First of all, it should be clarified that the figures of concentration–deconcentration represented something more than a simple movement of the spatial contraction–expansion of the metropolitan areas. It is true that the growing number of commuters to work has caused an expansion of the functional urban regions, with the addition of external rings formed by smaller towns with rapid demographic growth. This has emphasised the effects of counter-urbanisation in its formal definition of a higher percentage growth in small centres than in major cities. But this effect explains only part of the problem. In fact, especially in the 1970s, there was a generalised growth of minor urban systems not contiguous to the major metropolitan areas which, at the same time, were stagnating or even losing inhabitants.

Long-range deconcentration of this kind cannot be attributed to individual residential choices in favour of better housing conditions offered by small towns. This may be true for pensioners, but the strongest component of the phenomenon in the 1970s should be linked to the reduction in the number of jobs in the major cities and the growth in employment in the small and medium-sized urban centres. In particular, counter-urbanisation reveals itself as a spatial response to market, organisational and technological changes characteristic of the post-industrial (or post-Fordist) phase. In this phase, the smaller peripheral urban systems have been able to extend their range of externalities and become competitive with the metropolitan areas because of the convergence of various factors. These factors have reduced the monopoly exercised by the latter over the 'location market', with the result that a vast range of activities, both in industry and services, have been able to locate outside them.

In reality, some of the required pre-conditions for the start-up of this process had already been achieved in previous decades as a consequence of welfare policies that had improved the accessibility of infrastructures and collective services (schools, hospitals, etc.) in the small and medium-sized towns. But these have become externalities that could be enjoyed only with the rise, starting in the 1970s, of new forms of company organisation (deverticalisation, automation, spatial expansion of the network of suppliers, 'flexible specialisation', etc.), made possible by technological innovations in the fields of IT, telecommunications, logistics and high-speed transport.

This set of transformations has produced the current phase of 'urban transition', characterised by changes in the urban economic base, employment, demographic and social composition, in the forms of representation and institutional government and, parallel to these, in the spatial forms of urbanisation. These forms have thus offered much evidence of the processes underway, while not being able to explain them autonomously nor represent them completely.

This is also true for the more recent phases. Since the 1980s, a new movement of metropolitan concentration has emerged ('decentralised concentration') which is distinguished from the previous one by its scale and spatial form. In this, the growth in population and the location of activities become more geographically selective and tend to be distributed in peri-urban settlement patterns, which envelop the old metropolitan agglomerations and conurbations for tens of kilometres, or which develop along corridors that connect them. The new peri-urban metropolis no longer grows in compact areal forms but in wide-mesh networks and tends to spread in a way that invades and reduces the open spaces of the countryside, without, however, eliminating them.

The urban deconcentration recorded in Europe in the last thirty years is not thus reduced to a simple cyclical oscillation, destined to be reabsorbed by a new phase of concentration. This is a rather more complex phenomenon, operating on the global scale in all industrialised countries. It can be depicted as the result of two movements: one, more massive and visible, of deconcentration and another, more selective and qualitative, of strong concentration. On the scale of the individual urban system, the first prevails quantitatively over the second, while on the macro-regional scale the second takes its revenge, even if indirectly. On this scale, deconcentration appears, in fact, as a process of relative concentration in peri-urban belts around the few great metropolitan poles and the axes that connect them. Corresponding to these new spatial configurations are structural-physical, economic-functional, social- and political-institutional transformations which seem destined to produce new urban forms and types.

Types of city in the urban transition

The idea that the cities of Europe might constitute a single system is relatively recent and still remains largely to be demonstrated. It is a

Table 2.1 Some hierarchical-function types of European cities

Hierarchy levels	Brunet (1989)[a]	Kunzmann and Wegener (1991)[a]	Sallez and Verot (1991)[b]	Paris Group Cattan, Pumain, Rozenblat and Saint-Julien (1994)[a]	BfLR (1994)[c]	Dimensional Hierarchy (Brunet 1989) (population, thousand)[a]
I	Class 1 (2): London, Paris	Global metropolis (2): London, Paris	European capital cities (2): London, Paris	International dominant metropolises (2): London, Paris	City regions of world-wide importance (2): London, Paris, (Moscow)	Class 0: 6,400 (2): London, Paris
II	Class 2 (1): Milan Class 3 (7): Madrid, Munich, Frankfurt, Rome, Brussels, Barcelona, Amsterdam	Conurbation of European importance (6): Liverpool–Manchester–Leeds, Randstad, Ruhr, Rhein–Main, Copenhagen–Malmö	Major European cities: Ex: Berlin, Brussels, Barcelona, Athens, Munich, Frankfurt, Milan, Rome, Amsterdam	Specialised international metropolises (12): Amsterdam, Hamburg, Berlin, Copenhagen, Brussels, Munich, Düsseldorf, Strasburg, Frankfurt, Zurich, Geneva, Vienna	City regions of international importance (1): Brussels, Randstad, Ruhr, Rhein–Main, Berlin, Wien–Bratislava, Madrid, Rome, (St Petersburg), (Budapest), (Athens), (Istanbul)	Class 1: 3,200–6,400 (2): Madrid, Milan
III	Class 4 (11): Manchester, Berlin, Hamburg, Stuttgart, Copenhagen, Athens, Rotterdam, Zurich, Turin, Lyons, Geneva	Eurometropolis (12): Madrid, Barcelona, Lyons, Birmingham, Brussels, Hamburg, Berlin, Munich, Vienna, Milan, Rome, Athens	Eurocities: Ex: Glasgow, Birmingham, Seville, Lisbon, Copenhagen, Cologne, Rotterdam, Lyons, Hamburg, Marseilles, Vienna, Florence, Naples	Regional metropolises with strong international links (24): Ex: Manchester, Bristol, Lyons, Marseilles, Cologne, Stuttgart, Luxemburg, Rotterdam, Milan, Rome, Barcelona, Madrid, Basel	City regions of European importance: Ex: Manchester, Birmingham, Oslo, Stockholm, Helsinki, Copenhagen	Class 2: 1,600–3,000 (12): Ex: Lisbon, Glasgow, Manchester, Birmingham, Barcelona, Turin, Rome, Hamburg, Berlin, Stuttgart, Munich, Athens

Table 2.1 (cont.)

Hierarchy levels	Brunet (1989)[a]	Kunzmann and Wegener 1991[a]	Sallez and Verot (1991)[b]	Paris Group Cattan, Pumain, Rozenblat and Saint-Julien (1994)[a]	BfLR (1994)[c]	Dimensional Hierarchy (Brunet 1989) (population, thousand)[a]
IV	Class 5 (21): Ex: Vienna, Glasgow, Cologne, Basel, Utrecht, Marseilles, Bologna, Seville	Cities of European importance (17): Ex: Lisbon, Valencia, Grenoble, Zurich, Naples, Stuttgart, Glasgow, Salonika	Specialised European cities: Ex: Dortmund, Bilbao, St Etienne	Peripheral regional metropolises with limited international links (13) Ex: Southampton, Nantes, Belfast, Athens, Valencia, Lisbon, Granada	City regions of predominantly national importance	Class 3: 800–1,600 (21): Ex: Valencia, Marseilles, Naples, Liverpool, Cologne, Vienna
V and below	Class 6 (23): Ex: Bristol, Montpellier, Nuremberg, Liège, Bari Class 7 (40): Ex: Newcastle, Lausanne, Salonika, Rouen, Padua, Dortmund Class 8 (55): Ex: Coventry, Le Havre, Duisburg, Graz, Oviedo, Brescia		–Emerging high-tech cities: Ex: Cambridge, Montpellier –Subcontracting cities –Services and administrative cities (former country towns)	Regional metropolises with limited and very specialised international links (21): Ex: Cardiff, St Etienne, Kiel, Murcia, Bari	City regions of predominantly regional importance	Class 4 and 5: 200–800 (128)

[a] Europe of 12. Switzerland and Austria
[b] Europe of 12. Selection of examples only
[c] Europe, including western part of Russia. In brackets: 'potential' belonging to the class.

more or less implicit hypothesis that we find at the basis of many studies conducted in the past twenty-five years and which, whether right or wrong, has had the merit of encouraging a systematic exploration of the urban phenomenon on the European scale or, more often, in the more restricted area of the European Community, now the European Union.

One initial result of these studies is the definition of types of cities according to their functions, analysed both from the static point of view of endowments and the consequent hierarchies, and from the standpoint of their recent transformations, especially in terms of quality.

Studies of functional hierarchies are the most numerous. Table 2.1 presents a selection. In some of them, such as the one by R. Brunet, the hierarchical position of the cities is based almost purely on the level of functional features present in each of them. In others, such as that of the PARIS research group, a correspondence is established between the size and specialisation of the functional endowment and the range of territorial influences, following the principles of Christallerian theory. In these studies, the most widely used indicators, which carry most weight in the definition of the typologies, concern the international functions and connections, the higher tertiary functions (business services, research) and high-tech activities.

As can be seen in table 2.1, the results of the studies coincide only for the highest level, that of the 'global cities' (London and Paris). At the second level, that of cities of European importance, there are great divergences: for example, Manchester, Liverpool and Leeds are on the second level only if considered as a single system. In addition, there are particularly sharp divergences in the position of national capitals: Athens, for instance, can be found at the second, third or fourth level.

In the end, when compared, these hierarchical-functional classifications do not give much better results than those obtained by simply considering the demographic size of the cities (see the last column of the table). This depends both on the different choice of indicators and the weights attributed to them, and on the fact that the Christallerian hierarchical model, to which these classifications make implicit reference, is, as has been said, not very representative of the European situation today, even though it can still be found within each single country. On the European scale, all the cities of the levels considered in the table have supranational ranges of action, and what qualifies them hierarchically is this type of relation rather than their position in a system of regional relations. Therefore, these studies do not identify, either a unique system or well-defined hierarchical classifications. They do, however, indicate the separation of the level of 'European' metropolises (such as Randstad, Copenhagen, Stockholm, Berlin, Hamburg,

Brussels, Frankfurt, Vienna, Munich, Zurich, Budapest, Madrid, Barcelona, Milan, Rome), which occupy a privileged position, from the cities of national and regional importance, situated on the two lower levels (such as Dublin, Glasgow, Manchester, Nuremberg, Leipzig, Lille, Marseilles, Seville, Lisbon, Turin, Naples, etc.).

A very numerous group of cities (about fifty) thus stands out. With the unification of European space and the globalisation of markets, they have lost the 'monopolistic' advantages connected to their role as regional metropolises and in many cases have undergone considerable downsizing of their industrial production base. The most interesting current differences between European cities concern the forms of restructuring as a consequence of the growing openness of regional and national spaces. These differences have been highlighted in various ways. They may be underlined indirectly by the positive or negative deviation in their functional hierarchical rank compared to the demographic rank (Brunet 1989), or they can be measured with indicators of temporal variations. For this purpose, the simple demographic variations are of no use, in fact they are confusing, because, as has been seen, functional upgrading and demographic growth do not usually correspond. On the contrary, demographic variations are negatively correlated with size (the largest cities grow less) and are influenced by geographical position (Cheshire 1995; Cattan *et al.* 1994). It is worth, instead, considering the individual strategic functions, for example the presence of the headquarters of major corporations or advanced services companies, as M. Meijer does (1993), thus illustrating the upgrading of the level of European metropolises to the detriment of the immediately lower urban levels.

Fully fledged typologies derive from the consideration of a series of factors that describe the economic transition. Kunzmann and Wegener (1991) identify eleven types, of which five have positive growth and improvement trends: *global, technological and tertiary cities, new cities, monofunctional satellites* and *tourist–cultural cities;* four are in decline or with negative structural transformations: *industrial cities in crisis, ports, in growth without development* and *small cities and rural centres;* and two have differing dynamics according to the situation: *company towns,* and *frontier* and *gateway cities.*

Cheshire (1993), cross-referencing an economic performance indicator regarding the period 1971–88 with demographic variation produces four groups:

(1) good performance and demographic growth: above all medium-sized French cities like Grenoble and Montpellier;

(2) good performance and demographic decline: many cities in the central *dorsale*, from Amsterdam to Cologne, Frankfurt and Milan;
(3) demographic growth and economic crisis: many 'peripheral' cities such as Dublin, Palermo and Marseilles;
(4) demographic decline and economic crisis: above all old industrial cities and ports like Liverpool, Glasgow, Charleroi, Newcastle, Duisberg and Essen.

Conti and Spriano (1990) examined forty-eight major European agglomerations on the basis of about thirty indicators concerning the industrial production base, commercial, financial, management and research functions, infrastructural endowment, demographic dynamic and social composition. Using multivariate statistical analysis, the classes shown in figure 2.3 were obtained. Conti (1993) then extended this analysis to the cities of Eastern Europe.

Particular attention was paid to international functions. Soldatos (1991) introduced the distinction between *city-place* and *city-actor*. The former limits itself to receiving and offering passive support for institutions and international flows, regulated externally. The latter has an active social milieu which supports and organises locally international activities (even if of external origin) and produces and exports quality goods and services. De Lavergne and Mollet (1991) classify the cities into four groups according to the prevalence of these criteria: *economic internationalisation* (Bilbao, Lille, Turin), *services* (Madrid, Montpellier, Copenhagen), *infrastructural position* (Trieste), *attraction of specific flows* (Edinburgh, Lisbon). The authors also distinguish between cities that develop *explicit strategies,* i.e. coherent policies for the international development of the city and those whose *strategies are implicit or absent*, in that they derive from the sum of a number of public and private initiatives poorly connected to each other.

Another research project looked at eighteen major non-capital European cities (Bonneville *et al.* 1993). This distinguishes three large groups of international cities: those with an internationalised production base (e.g. Stuttgart), those which act as interface between the world economy and the regions (e.g. Milan) and those that play functions of financial or political regulation on the international scale (e.g. Frankfurt, Geneva).

These various classifications as a whole reflect transformation processes that act on the global scale. The resistance that they encounter and the responses that they receive in the various urban situations generate the broad spectrum of types just listed which, however, especially since the late 1980s, converge towards a form of dualism. On the one hand, we have the trajectories of growth, functional qualification and

1. GLOBAL CITIES

Pure	London	Brussels Amsterdam	Rome Copenhagen
Completed	Paris	Frankfurt	Milan

2. CITIES IN POSITIVE INDUSTRIAL AND TECHNOLOGICAL TRANSITION

Pure	Stuttgart	Turin
Complete	Munich Nuremberg Düsseldorf Cologne Strasburg Hanover	
Incomplete	Essen Bologna Lyons Grenoble Bochum Dortmund	Bordeaux Toulouse Duisburg

3. CITIES IN NEGATIVE INDUSTRIAL TRANSITION

Highly tertiarised	Dublin Liège	Utrecht The Hague Rotterdam
Maritime tradition		Marseilles Genoa Antwerp

4. CITIES IN STRUCTURAL CRISIS

Urban obsolescence	Naples Edinburgh Glasgow Manchester Lille
Industrial restructuring	Birmingham Bristol Nantes Nancy

Figure 2.3 Dynamic-functional classification of forty-eight major European urban agglomerations (Conti and Spriano 1990).

social polarisation that characterise a few major metropolitan systems at the global or at least European level. On the other, there are the large number of cities, mainly small and medium-sized, but sometimes also large, which experience the urban transition as a permanent crisis, as a slowdown of growth and functional qualification, or even as successful sectorial specialisation that is increasingly controlled from outside and subject to the fluctuations of the global market. More generally, it can be said that the global network connections now operate directly in all cities, large and small. This challenges traditional identities and drives the most enterprising local elites to replace them with new pro-active identities, in which local social cohesion is a condition for openness to higher spatial scales.

Cities as local territorial systems and as nodes in global networks

The trends illustrated so far allow us to outline two main levels of spatial representation of the city. One is the *local* level of the individual city, in which space means *proximity* and presumes interaction between actors (or potential actors), in the presence of a given set of resources and a specific local milieu. Another level is the supralocal one, tendentially *global* (in our case European), where space is given by the networks of flows and material and 'immaterial' relations that link the various cities together, *irrespective of the distance between them*. These two types of space – the physical–territorial one of interactions of proximity, and the virtual or topological one of flows, i.e. interaction at a distance – are very different on the logical and conceptual plane and, in some ways, in contrast with each other. But this does not mean that the urban phenomena that occur at the local level are incompatible with those on the global level. One might be led to think this if one is not able to distinguish the simplified form of spatial representations from the complex nature of the urban phenomenon, transforming purely conceptual contrasts – such as those between local identity and global standardisation, between places and non-places, between endogenous and exogenous urban development – into the terrain for ideological conflict.

As these new fundamentalisms are at least generated and supported by the hidden persuasion exercised by spatial images, it should be stated clearly that, although the cities can be depicted conventionally both as communities rooted in the territory, and as nodes of global flows, they are neither one thing nor the other, but a third kind, in which the two above-mentioned spatial forms offer partial images.

The city of today cannot be thought of only on the local scale as a stable, organic entity, formed by a physical 'body' (*urbs*) and by an organisational 'mind' (*civitas*), which make them capable of strategic projects and action. This, which was considered as a natural fact in the past, has become something that has to be designed and constructed. In fact, urban diffusion not only means the loss of the borders of the city's 'body', but also that the 'central' functions which once characterised the heart of the individual cities are being redistributed in space as nodes of a network, where centre and periphery mingle. And even where the city, so extended and fragmented, conserves formal recognisability, it is doubtful whether the individual and collective actors that compose it constitute a cohesive group just because they co-exist in one place. These networks, which tend to be global, cross the cities and connect their actors together at a distance, weakening the traditional bonds of internal cohesion, founded on physical proximity.

However, if the city-actor rooted in its territory is no longer a given fact to be taken for granted, the typological analyses examined above demonstrate how many European cities still continue, perhaps more than before, to represent themselves and behave as 'strong' collective actors in the global spaces of network competition and co-operation. Recent transformations have thus not eliminated urban territoriality, but have modified its substance, accentuating its role, shifting it from passive to dynamic, from a simple product of long historical duration to a product of local organisation, from a value in use enjoyable within a limited geographical area to (almost) value in exchange, from a sort of 'legacy' to be preserved to 'risk capital' to be staked in global competition.

An image thus begins to form of the city as a 'node' of global networks, where local identity and the urban territory, as a stratified deposit of natural and cultural assets, no longer have a value for what they are but for what they become in the processes of valorisation. The partial truth contained in this image is that the city as local society is no longer identifiable for its stable embeddedness in a given territorial milieu. It is instead a changing connective configuration with variable actors which can be thought of as 'nodes' of local and global networks. And it is thanks to these actors that the networks meet, interact and connect in the city. And it is for this that, in a virtual space of networks and flows, the cities as territorial 'bodies' and as social aggregates continue to exist and to play an essential role, in precisely those processes of globalisation that would seem destined to destroy their identities.

Something important is thus changing. For the global networks, the urban territorial milieu does not offer real *roots* but simple *anchors* (Veltz 1996). It cannot therefore be either defined or described as an objective entity, but can only be grasped in the moment when it offers 'grips' for these anchors. The new urban territoriality is something that can be observed empirically only through the effects that it produces. It is a conceptual image that allows us to understand today's protagonism of the cities and their nature as attractors–connectors of global networks: thus, the formation of local social networks around projects of valorisation of the resources typical of a local context, not only those aimed at the market, but also at local social circuits understood as a shared commodity and autonomous value.

The disappearance of obstacles and friction to the world-wide deployment of market forces has the secondary effect of creating uniformity in certain places (those that M. Augé (1992), calls *non-places*: shopping malls, airports, etc.). But this does not mean that the same thing happens between *territorial systems*, including cities. At this more aggregated level, the principal effect of globalisation is not standardisation but the multiplication of *connections*. The cities compete with each other to assure themselves these connections, i.e. to attract the 'nodes' of the global networks made up of the major public and private transnational organisations, by financial markets and specialised services, scientific and technological co-operation, media and so on. But at the same time, globalisation generates resistance, exclusions, conflicts. By transforming the specific conditions and resources of the various urban and regional milieux into *competitive advantages*, it makes them act as powerful factors of local diversification.

For their part, the global networks see the anchoring of their 'nodes' in diversified local systems as opportunities capable of increasing their competitiveness on the global scale, both for the possibilities of exploiting the particular local resources and conditions in the multilocated and flexible organisation of production, and as access to differentiated segments of demand, with feedback effects on product and organisational innovation. It can also happen that those excluded, or self-excluded, from these processes produce forms of social interaction that then tend to be 'colonised' by the global organisations. One example is voluntary social work and the 'third sector'.

In recent years, the author has attempted with a number of colleagues to represent together the local (territorial) space of the city and the global (network) space by adopting the theories of complexity. In particular, the theory of auto-poiesis offers fertile logical terrain for this coupling (Dematteis 1991; Conti 1993; Conti, Dematteis and Emanuel 1995). In this, the individual cities are considered as self-organised 'local territorial systems' that interact with an 'environment' formed by similar urban systems and by global network organisations. From this standpoint, the cities can thus be imagined as nodes of global networks while as individual self-organised territorial systems they should be seen as local networks of local and global actors, immersed in a given urban–

territorial milieu. From the interconnection of these local networks stems the *structure* of the city system, which changes in time, as a response to external disturbances and stimuli. The structure is found, therefore, both in the internal cohesion of actors ('operational closure'), and in the functional openness of the local urban system towards the outside ('structural coupling' with other systems). Internal cohesion and external openness do not rule each other out reciprocally. On the contrary, long-lasting urban development presumes situations in which good integration of the local social networks allows the city not only to attract a higher number of flows and nodes belonging to the global networks, but also to produce the forms of internal social interaction needed to govern relations with the outside to its own advantage.

City networks

The interactions that connect together the various 'nodes' of the global networks (companies, financial, commercial, political and cultural institutions, etc.) belong to the virtual space of flows. On a geographical map they tend to converge and to centre on a relatively limited number of places, corresponding to the various urban territorial systems. The cities, because of the fact that they are linked to each other by these bands of flows and interactions, can thus be considered as complex nodes, formed, in other words, by a number of interconnected nodes of complex networks, in their turn formed by bands of numerous networks (Gottmann 1991). In this sense, we talk of *urban networks*. The same expression is increasingly used in the stricter and more technical sense of voluntary networks of alliance and co-operation between cities *(urban networking*: De Lavergne and Mollet 1991).

The studies of European urban networks have used both direct indicators, which measure flows and interactions, and indirect ones, which consider the excesses/shortages of individual cities' functional endowments as sensors of their reciprocal relations (of command, exchange, complementarity, and thus presumed circulation of people, goods, capital and information), or by examining the configurations of the physical transport and communications networks that join them.

The direct measurement of flows is hindered by the scarcity of statistical data on an urban basis. The most accessible sources on the European scale concern air and rail traffic. On this basis, N. Cattan has reconstructed the hierarchical relations between the main European cities (Cattan *et al.* 1994). Other relations of domination/dependency can be deduced from the geography of multilocated corporations, and in particular from the location of headquarters and branches controlled by them (Pumain and Rozenblat 1993).

From the physical road and rail network that links the cities, a hierarchy can be drawn up based on the node values that, especially for the high-speed railways and motorways, are usually associated with values of urban centrality. Again starting from the transport networks, the cities can be classified according to their reciprocal accessibility, highlighting above all the strong centre–periphery relations still present in Europe.

No rigorous analytical research has been conducted on the model of central places in Europe. The results would probably be disappointing, in that the model is based on relations of proximity that have existed for centuries within the individual countries and rather little across political borders. The opening of the single European market is too recent to have had noticeable effects on the restructuring of cross-border relations of proximity. Furthermore, in a space of flows and interactions less and less conditioned by the 'friction of distance', the hierarchical relations between cities now display different geometries. The model of central places is used, instead, in a 'negative sense' to highlight the deviations compared to expectations and in particular the non-hierarchical but complementary relations between centres. While the former give place to star- or tree-shaped configurations and thus to non-interconnected figures, the latter give rise to highly interconnected figures. In figure 2.4, showing the central and western part of the Po valley (Piedmont and Lombardy), these diverse forms are present. The star and tree-shaped forms prevail in the southern agricultural area, while the interconnected configurations characterise the industrial–urban belt at the foot of the mountains, with a peak north of Milan.

Similar phenomena are found in the most developed and densely inhabited parts of Europe (Batten 1995). The most recent cartographic representations of the urban phenomenon reflect, in the symbols adopted, this shift of the urban shape towards network structures. These offer a sort of functional x-ray of the physical forms of peri-urbanisation and the diffuse city as illustrated above.

Models and scenarios for the European urban network

The spatial configurations of the European urban network can be summarised as indicators of the transformations in progress. To this purpose, it is worth going back to the schemes in figure 2.2. Type A represents the urban network as a hierarchy system on several levels formed by centres that tend to be equidistant: this is thus a territorially balanced system, as the hierarchical relations between cities are regulated by their

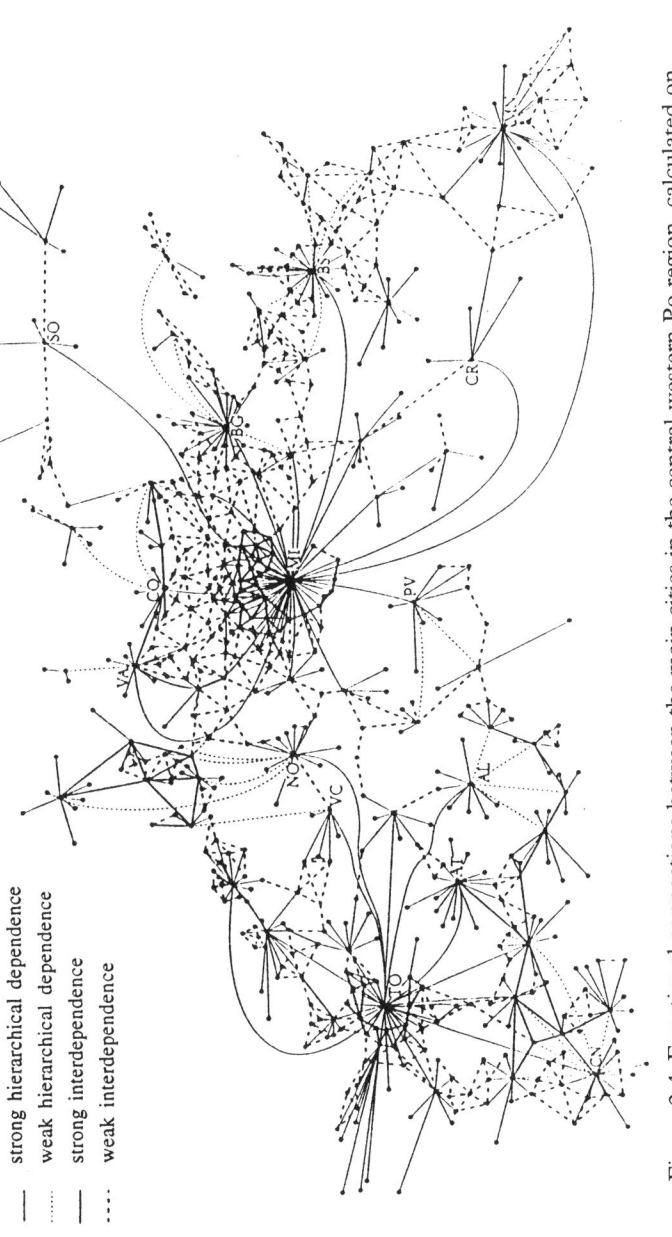

Relations of:
— strong hierarchical dependence
⋯ weak hierarchical dependence
— strong interdependence
- - - weak interdependence

Figure 2.4 Functional connections between the main cities in the central-western Po region, calculated on the basis of distances and the cities' endowment of services (Emanuel 1988).

reciprocal distances, according to what occurs in the Christaller model. The type B configurations also represent a hierarchy on many levels, but not spatially balanced, because the rank of the city diminishes, shifting from the central region to the peripheries. Finally, the type C configurations correspond to a system that is also hierarchical, but in which the relations between the individual cities can be both vertical (hierarchical) and horizontal (of functional complementarity), for which each city of a lower level can have horizontal relations with a number of cities of any level and vertical relations with cities of a higher rank, independently of geographical position and distance.

These spatial configurations represent different economic–social structures. The first corresponds to a situation of equilibrium typical of a pre-industrial market economy. The second is typical of the 'Fordist' industrial system of the first half of our century, characterised by economies of scale and agglomeration, which create polarisation effects at the various territorial levels. Configuration C corresponds to the more recent information economy, characterised by this multiplication of distant connections between actors, both vertical and horizontal.

As these three models have succeeded each other in the last two centuries, in very different periods and ways in the various parts of the continent, today's network of European cities sees them partially overlapping. On the continental scale, type B (centre–periphery) is still dominant. All recent studies recognise a core zone, that lies along the Rhine axis, where the highest density of top-rank cities is located, variously defined by the different authors: from the more limited 'golden triangle' of Brussels–Amsterdam–Frankfurt (Cheshire *et al.* 1989) to the vaster 'European megalopolis' of J. Gottmann (1976) and the *dorsale* (also known as the 'blue banana') of the French DATAR research group (Brunet 1989).

This concentration shows a central area in which 53 per cent of the European-level metropolises are located, with a geographical density six times higher than that of the most peripheral area and a ratio of urban population 2.2 times higher (Dematteis 1996). The highest levels of network interconnection envisaged by model C (Cattan *et al.* 1994; Lutter 1994) are recorded within this central region, while in the 'European periphery' the permanence of model A is found, in addition to the rarity of higher ranked cities. Usually here, the connection between the lower urban levels and the European metropolitan level is still mediated by one or more urban levels of regional or national range.

Nowadays, therefore, we can talk of a full-scale European network of cities only as far as the upper levels of the 'global cities' and 'European

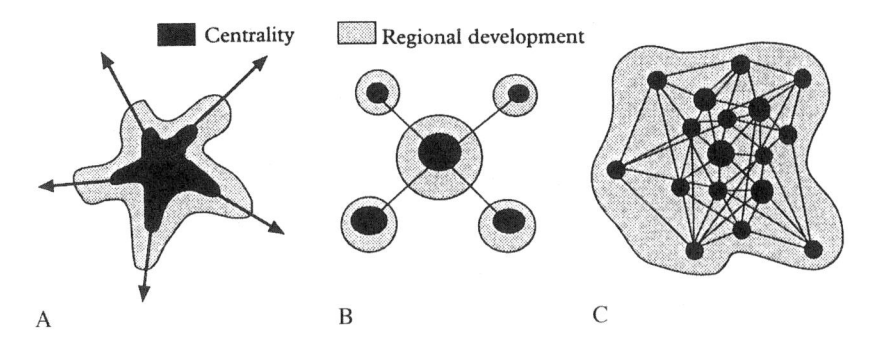

Figure 2.5 Three schematic models of the evolution of European urban centrality (in black) and its effects on regional development (in grey): A = semi-peripheral extension of the 'core'; B = hierarchical decentralisation; C = distributed metropolitan centrality.

metropolises' are concerned. Below them, if we exclude the central continental regions – where numerous small and medium-sized cities are already integrated into the European network – a fragmentation into national and regional networks inherited from the past usually continues (Rozenblat 1996). It can also be observed that there exists an evident spatial correlation between the degree of interconnection of the urban networks and the indicators of regional development and well-being. They show higher values in the European urban core and minimum ones in the 'peripheries' such as the Mediterranean, where the urban network is rarefied and discontinuous. This correlation assumes a geopolitical significance if we consider that the strongest European states are also those that include parts of the *dorsale* or the continental megalopolis. The urban network should thus be considered as a strategic component of the European Union's policies of regional cohesion and re-equilibrium. The recipes proposed so far can be summarised in the schemes in figure 2.5.

In the first (A), the problem is solved through the expansion of the central area, acting essentially on the land transport infrastructures, in particular motorways and high-speed railways. As recent studies have demonstrated (Lutter 1994), this solution is destined to produce further imbalance, as it strengthens the semi-central intermediate zones to the detriment of the peripheral ones. Scheme B, referring back to the Christallerian model, envisages the hierarchical decentralisation of functions from the central area to the peripheral metropolises, which is destined to improve the economic and social situation of their surrounding regions. This improvement will still remain, however, inferior to that of the central regions, as learnt from the analogous policy of balanced regional metropolises as practised in France in the 1960s and 1970s.

Scenario C appears in theory as the most effective, even if it is the most difficult to achieve. In contrast to the other two, it envisages spatial transformations (extension to the peripheries of the central area's levels of centrality and development) through intervention that does not act directly on physical space but on the actors in local development. These are thus processes and policies that cannot be represented with normal geo-cartographic images. They would demand passages in scale – from the local to the European – and need to be registered at the same time in the spaces of proximity, in the virtual ones of network interaction and in those of the collective imagination of local actors. All of this is difficult to translate into images as strong as those which support policies of direct intervention. Neither are there strong interests that push in this direction, as the policies to support local urban development are far from mobilising the millions of dollars involved in infrastructural works. Without ruling out the effectiveness of a European policy of incentives and of redistribution of metropolitan urban functions (which in any case would still largely need to be invented), the local level appears in any case the one from which to start to create these new images. On the European level, the one that gets closest is the 'grape' proposed by Kunzmann and Wegener (1991) as an alternative to the 'banana' of the megalopolis.

Beyond appearances: the problems of the new metropolitan realities

This rapid critical review of European urban-spatial imagery allows us to focus on a number of problems. First of all, the disproportion between the wealth of production of spatial images, examined only in part here, and the modest results that these images have produced on the theoretical level and in technical-operational terms, seems evident. As has been seen, the attempts at theoretical formalisation (city life cycle, Christallerian hierarchies and similar theories) are based on over-simplifying foundations, which are unsuitable for explaining a complex and dynamic reality like the cities today. This theoretical and analytical weakness is reflected in the absence or near-absence of technical content in the planning and urban policy documents produced by the EU and by the governments of its member states. The ample recourse to spatial language in these documents is largely rhetorical, not only in the general sense that the language of geographical space is in any case a sort of metaphorical code for talking about social issues and implicit political projects, but also in the stricter sense of an artifice designed to persuade. Representation of the city with dotted symbols or micro-areas suggests the idea that they are still today unitary and cohesive entities; the distances that separate them evade more complex discussions on their reciprocal relations; the urban networks give a stable form to what is in reality destabilising; the typologies of city oversimplify the variety and conflict of responses that each of them gives to the processes of globalisation; distributed on a map, they bring out spatial configurations that impose themselves as the only ones pertinent to urban policies.

The basis of this use and abuse of images and spatial language is, perhaps, the implicit need to create order and stability in a world in which the acceleration of change and the evident inability to govern it generate insecurity and uncertainty in the future. This need should certainly not be underestimated. But it is doubtful that the best way to satisfy it is to depict the European urban network as cohesive and stable when in reality, as has been seen, it is fragmented and has many nodes in crisis. This is even more true when one neglects (and does not investigate and represent) what should and perhaps could really be stable and which concerns above all the geographical scale of everyday life. For instance, the forms of 'local' social interaction should be considered which, even more than fighting exclusion, can guarantee inclusion: the practices and structures that are not just simple responses of adaptation to the destabilising stimuli of globalisation; the ones that respond to the needs of life well before the diktats of the global economy. A geography of these truly 'stabilising' facts and projects is lacking both on the real scale of the phenomenon and on the higher national and European scales. And this happens despite the fact that its images could be equally effective and formative to those of the functional networks of cities and perhaps more useful in terms of reducing instability and insecurity.

Beyond their rhetorical use, urban spatial images have also, as has been said, a not negligible heuristic role: connecting together a quantity of 'superficial' evidence, they mark changes and problems, suggest hypotheses, challenge consolidated images. One doubt concerns the possibility of making the model of the political city-actor correspond to well-defined territorial entities. At the higher level, that of the global cities and the great European metropolises, the spatial form of the city appears increasingly extensive and less and less compact. As we have seen, peri-urbanisation does not consist only in a redistribution of residences. The new network forms of settlement which surround the metropolitan nebulas are themselves articulated around nodes and nuclei that have relations of complementarity with each other and with the old centres. On this scale, centrality, no less than social classes and groups, is spatially fragmented, while functional cohesion is found at the level of the global networks. Each territorial–urban 'fragment' is connected directly to them, i.e. no longer passing through the connective mediation of a metropolitan centre capable of giving unity to the intermediate (regional and national) territorial levels, as once happened.

It thus seems that a vast, articulated and centrifugal territorial aggregation such as a large metropolis today can be traced back to the city-actor model (or constellation of city-actors) only as far as the local connective relations based on relations of proximity are concerned, such as the management of infrastructures and collective services. It is instead difficult to imagine a complete structuring (of a Weberian type) of local urban life and society corresponding to new metropolitan territorial forms. The nature of this difficulty needs to be examined further. Is it just a question of the size being too great or the degree of articulation too complex, that can be resolved by looking for new modes and techniques of governance? Or does the new territorial structure of the metropolis reflect a 'catastrophe' of a structural type, deriving from the definitive shifting of the development processes (and their governance) from the local–regional level to that of the global networks? And in this case, do the two levels, the local and the global, remain necessarily separate, or, although local society has lost total control of urban development, could it still aspire to a unitary structuring around weaker forms of governance, aimed at managing the interactions between the two levels – local and global – according to the scheme we have seen of 'structural coupling' between auto-poietic systems?

It does not seem that these questions can be avoided simply by turning to the lower urban levels, and in particular to the medium-sized cities, which appear to conserve better their role of local and regional territorial co-ordination. It is certainly a good choice to start from the analysis and comparison of simpler cases like these. But, in my opinion, this should be done to tackle the real problem of the metropolis better, and not to avoid it. In fact, the images of European urbanisation clearly indicate that the medium-sized cities do not escape (or will not escape for much longer) the 'catastrophe' of metropolitan globalisation. The crisis of economic and social restructuring that they are going through seems to be a symptom of a more general restructuring that will see them increasingly inserted as nodes (in the successful cases) or as decaying fragments (in the cases of decline) in a territorial mosaic which is appearing on the European scale. It repeats, on a large scale, the same processes of territorial fragmentation and of network connection/disconnection that can already be seen clearly in the regional spaces of metropolitan peri-urbanisation. If this is how things stand, the contrast between medium-sized cities as models of 'regional cities' and metropolises as 'global cities', while maintaining an element of truth, risks concentrating attention on transitory situations, destined to change rapidly in the years to come.

The types of European urban governance should, in my opinion, be able to refer to the different territorial forms (nodal, areal, nuclear-fragmented, network, etc.) that the urban phenomenon is assuming today, even independently of set dimensions, borders and geometries. The important thing is to recognise that each of these forms corresponds to frameworks of social interaction relating to everyday life, whose 'anchor points' are found in certain specific conditions of the local milieu. These environments seem to constitute today the only relatively stable territorial supports for local collective action capable of resisting the destabilising impetus transmitted from the global networks and interacting with them.

3 Segregation, class and politics in large cities

Edmond Preteceille

A long while ago, the city was where politics such as we still to a large extent know it today was invented. The mutual institution of city and citizen thus materialised there where population is at its densest. Further it came about through affirmation of the division between the citizens and the rest, between slaves and foreigners. Subsequently, the emergent nation-state widened the territory of citizenship and correspondingly reduced the political importance of cities, although the larger ones – those most often designated as capital cities – continued to provide the arena for intense political life where political power was concentrated, exercised and contested. Hence the rise in the economic and political importance of cities went hand in hand.

Today, the growing part played by supranational institutions, in Europe especially, and the development of economic globalisation in turn are reducing the role of states and the significance of national territory as a political arena. Scepticism over the coming emergence of European citizenship has the effect of producing a counter-interest in the renewed significance of infranational political areas, cities and regions appearing as rival candidates.

In this respect, large cities are in a paradoxical situation. On the one hand, if globalisation results in cities and even regions constituting territorial divisions which coincide less and less with economic flows and relations, the various readings that are put upon this economic transformation concur in recognising an advantage and a specific position for the larger cities, namely the thesis of the global city put forward by S. Sassen (1991) or that of the insuring function of the great metropolis advanced by P. Veltz (1996). On the other hand, their institutional complexity, their size in terms of geography and population and the intense social divisions they exhibit would appear to make a relatively unified and effective political crystallisation of economic potential more problematic than in smaller cities. It is the latter point which will be explored in the present chapter. What are the political consequences of the social divisions that characterises major cities? Can such divisions account for the problems they have in emerging as political actors with an important role to play in the new forms of 'governance'?

And indeed the larger cities manifest profound social divisions, which seem to have become accentuated with recent economic and urban transformations. In Europe, government, local representatives and the media both recognise and deplore the situation, and policies are drawn up and promulgated to attenuate such divisions, especially the social violence considered as their consequence. There is nothing new about the existence of such divisions and the studies of them carried out by Engels and others formed part of the pre-history of urban sociology before constituting one of the special subjects of the Chicago School. But the views of the large city have undergone change. In the twentieth century and especially during the prolonged expansion following the second world war, large European cities were perceived as places of progress, culture and access to modernity, as focal points in the 'civilising process'. Social divisions were destined to become blurred, and the production of urban facilities for suburban developments, the provision of social housing, the gradual disappearance of shanty towns and slums, the growth of consumerism and car-ownership were clear manifestations of this.

However, for the last fifteen years or so the fear of social breakdown has reappeared in the way the great metropolis is represented, outbreaks of violence have borne witness to the inequalities, the discord and the sense of exclusion that brand certain suburbs. A North American, even a Third World, contagion is spoken of in connection with the larger European cities; and they are compared unfavourably with small and medium-sized cities, with their easy social relations and warm sense of identity.

How is one to account for these social divisions, their absolute and relative intensity and the polarities that seem to aggravate them? Do they imply an improbable reassertion of class conflict or its final eclipse and a new divisiveness in which social class is replaced by exclusion or ethnicity? And then how are the processes which produce such divisions to be analysed? How are the cross-effects of economic changes and labour markets, housing markets, individual attitudes and political and institutional practices at local, regional and national levels to be disentangled?

Finally, what are the consequences of these divisions and the processes that produce them? In particular, what are their effects on the development of social inequalities, on the relations between social groups and on the construction of social identities; and, further, on political practices and the capability of groups, of local communities, of cities too, to become political actors and so contribute in a significant way to the construction of their future?

Social divisions, ethnic divisions

What is known about social divisions in major European cities? Are there common features – or common differences – as compared with American or Third World cities? And are there any recent signs in such divisions to suggest convergence towards a single 'major city' model; or rather do specifically European features seem to be confirmed or to have become more pronounced?

The empiricist descendants of the Chicago School attempted in the 1960s and 1970s to compare the factors, forms and degrees of intensity of social segregation in cities: factorial analyses of sets of descriptive variables of social structure, which provided empirical evidence of the dominant pattern of social structure; calculations of the indices of segregation for the whole of a city as a basis for comparison; studies of conformity with spatial models, whether concentric (Burgess) or sector-based (Hoyt). But there is no purely empirical response to the question of knowing what the principal patterns of segregation are, and on this point comparisons between cities prove to be tricky to say the least.[1]

Patterns of segregation which have been the most subject to analysis have been those relating to class divisions or racial divisions, to which one might add religious denominations. These patterns will be discussed here, although research indicates others that merit attention, such as population differentiation by age and by size and structure of households, and gender relations.

The study of spatial division by social class raises two questions: that of the definition of these classes and that of the results observed, the one being not unconnected with the other, as has been remarked. Seventy-five years ago, the existence of major social classes defined by the division of labour would, rightly or wrongly, have raised barely a murmur from sociologists. Both M. Halbwachs and R. E. Park noted their significance in the differentiation of urban space, and Park was of the opinion that social class segregation was particularly strong in Europe, especially in London: 'In the older cities of Europe, where the process of segregation has gone farther, neighbourhood distinctions are likely to be more marked than they are in America' (Park 1925: 11): hardly a judgement to be expected from a contemporary European, or for that matter from an American.

Is social class still an appropriate category for the analysis of current urban segregation? Most sociologists agree in recognising that the view represented above, particularly in its hardened Marxist form that is due in fact more to Lenin and Stalin than to Marx himself, is at once too dogmatic and simplistic – assuming huge and stable social identities which would be adequate for an interpretation of the social differences observed – and also too static, being particularly ill suited to accounting for current social developments (the decline of blue collar workers, the progression of middle and higher-waged categories and of a tertiary proletariat) and for the transformation of the modes of constitution of social identities. But in my opinion the debate remains open as to the actual relevance of the Marxian schema itself – the hypothesis of the major formative effect of the division of labour and of the relations of production upon the definition of social divisions.

To oversimplify, at the present time two contrary theoretical positions can be distinguished. The first plays down the importance of economics in social stratification. I would list in no particular order, but without in any way assimilating them: the neo-Weberians such as P. Saunders (1986), who see decisive elements of social status in consumer patterns and house-ownership; the Pierre Bourdieu school with its insistence on 'educational capital', 'social capital' and symbolic domination; and those who draw arguments from the long-lasting – if not permanent – exclusion of part of the population from wage-labour in order to assert the growing importance of non-economic processes in the definition of social identities (cf. R. Castel 1995, for discussion of the crisis in the wage society, 'disaffiliation' and 'supernumeraries', or W. J. Wilson (1980; 1987) and the ongoing debate on the underclass in the United States).

Conversely, the other camp is inclined to see in the Fordist crisis, in the advance of neoliberalism and in globalisation, arguments that underline the growing pressure of capitalist production relations dominated by multinationals and financial markets on the way that the whole of social life is organised, cities included. Here I would mention many number of studies influenced by the theses of the regulation theory school (cf. Amin (1994)) and the work of Sassen, referred to above.

If one looks at available findings, it is immediately apparent that almost all recent empirical analyses of major cities[2] that examine the social breakdown of space according to characterisations of population referring to economic processes (such as occupation categories or 'catégories socio-professionelles' in France) give prominence – as indeed do earlier analyses – to a primary factor describing the opposing in distribution between higher categories and worker categories, in spite of the absolute and relative decline of workers.[3] It is also apparent that in cities like London, Paris and New York, which have an industrial past, today's predominantly working-class areas are to a large degree those that were so a century ago (probably rather less true in the case of New York where the transformation in the social use of space has occurred more rapidly).

However, the more pronounced form of segregation in these cities has to do not with the working but with the upper classes. This fact is largely overlooked in commonly held views where segregation is habitually associated with the poor. Of course, the deliberate choice on the part of privileged groups to be 'with their own kind', to have their own space and direct its use in accordance with their own values (Pinçon and Pinçon-Charlot 1989) is not on the face of it a social problem deserving of care and compassion. But this very marked spatial concentration appears to me to raise a number of problems which will be discussed later in this text.

Above and beyond these highly general common features, what evidence is there of significant differences between large cities in Europe and in North America? What signs are there of convergence in similarity of structure, as suggested by notions of the Americanisation of European societies or, in a different mode, by the theoretical model of the 'global city' (outlined by Friedmann 1982 and Friedmann and Wolff 1986) and developed more recently by S. Sassen (1991)? Furthermore, is there disparity or continuity between the larger cities in Europe and the others (the model of the global city makes the assumption of there being a widening gap, whereas traditional geographical models of cities in order of size take for granted a degree of continuity and a top-down spread of innovation from the larger cities)?

Mention has been made of the methodological difficulties of such comparisons (dissimilarity of statistical categories from one country to another, dissimilarity of spatial divisions even between cities). In an essay comparing London, Madrid, New York and Paris (Preteceille 1993), it was found that segregation seems more intense in European cities, which would show that R. E. Park was still right today. But the more comprehensive definition of higher categories in the United States (the variable 'occupation' in the census gives a rather wide definition for higher categories: 'managerial and professional speciality occupations') and the larger size of the spatial units that we have been able to use make this a questionable result. Besides, if one looks at the concentration of workers in predominantly working-class areas, one is inclined to conclude that there is less segregation in Paris than in New York, whereas in London it is higher – but there too, the findings are uncertain and provisional.

Comparison with Madrid shows more solid methodological differences pointing to more marked segregation there than in the three other cities, which again does little to corroborate contrast between European and American models. Conversely, studies undertaken in Athens by T. Maloutas show segregation there to be less marked than in the other major European cities, which should lessen the appeal of a more segregated 'Mediterranean' model inasmuch as Europe as a whole proves to be too diverse.

Nor is the comparison of the spatial organisation of segregation in different cities any more straightforward. Traditionally a contrast has been made between European cities, where inner city values are enhanced by the presence of the upper classes, and American cities, where the better-off prefer the suburbs leaving the poor in the inner city which has become gradually abandoned in the shifts of modernity. But here New York hardly conforms to the American model since certain sections of Manhattan are among the most up-market of the entire city. Likewise, Paris gives a different shade to the 'European model' since areas of upper-class concentration include eighty-four suburban *communes* together with twenty-seven of the eighty districts in the city itself (Preteceille 1993: 26). And London diverges even more, since if those at the top[4] of the social ladder tend to be concentrated in the west and north-west of the city, the social category next below them, which is less compact as a group but twice as large, is mainly located in the greater London outskirts.

If trying to compare the structuring of social divisions in major cities at a given moment in time presents problems, comparing trends is more delicate still. It is, however, worth noting a degree of convergence in findings by C. Hamnett (1995) on London, M. Sonobe and T. Machimura (1996) on Tokyo, and my own on Paris (Preteceille 1995). With the combined objective of testing the validity of the global city model in our respective cities we concluded, in the face of the hypothesis of dualisation, that there was a marked rise in the higher categories, a relatively less significant rise in middle intermediate categories, though it was still strong in absolute terms, and a drop among industrial workers as well as in office workers (which is, however, not the case in Tokyo). If it is true that a new tertiary proletariat whose jobs are insecure is partly taking the place of the industrial proletariat, it is not showing the significant growth projected by the model. If it is also true that inequalities in income are increasing and that the highest earners are seeing their proportion of total income increase further, there are, on the first findings of current surveys, no signs of strong impoverishment among the lowest wage-earners, nor indeed of any drop of income for intermediate categories. Hence, pending confirmation of these findings and of their validity in the case of household income, it would be wrong to characterise the impact of ongoing globalisation as an overall dualisation of the urban structure, echoing, in a sense, through the domination of the globalised tertiary industries, the binary class division in cities of nineteenth-century industrial capitalism. If it is further true that these economic activities are indeed the ones that have played the most dynamic role in the economic transformation of the Ile-de-France region and of its labour market, proper account should be given to the industrial activity still remaining there, much of it linked to high-tech sectors, as well as to essential public-service employment.

Thus, contrary to a commonly held notion and to assumptions made by a number of researchers, it is wrong to visualise the big city, and the Paris metropolis in particular, in the context of dualism, though one may well allow for limited trends of spatial dualisation. For instance, areas within the Ile-de-France where the higher social categories are already concentrated have become further specialised in this respect, as have two thirds of the *communes* with the highest working-class concentration. But 60 per cent of the region's population live in areas where trends are more complex and where there is no significant polarisation.

The greater Paris area then is no more than partially affected by spatial dualisation, nor does it have the greatest concentration of the very poor, contrary to the accepted view. The question has been touched on briefly as regards the overall trend in the active population. But in terms of space, even allowing for the partial polarisation mentioned, high rates of working class and/or poverty are not located in the Ile-de-France. In her social and professional typology covering the whole of France, N. Tabard has shown that the area of greatest working-class concentration in Paris[5] included a larger proportion of those in the higher categories than in the exclusive districts of many other French cities, and that the poorest *communes* in suburban Paris were only a little below the average. Likewise, the median situation of sectors in these *communes* targeted by urban policy, theoretically the poorest, is better than that of most comparable sectors in provincial cities.[6]

Certainly, social contrasts are at their most flagrant in the Ile-de-France, but this is due to the very high concentration of wealth, not to the absolute level of poverty. From the viewpoint of spatial inequality, the gap between targeted sectors and the others is far greater in the Paris region[7] because the average for the other sectors and *communes* stands out by reflecting the degree of high-income residents, just as the gap between departments displaying extremes of social structure (Paris and Seine–Saint-Denis) has grown wider because, if the higher categories have progressed in both, their progress in Paris has been much greater.[8]

Evidence is hard to come by in the case of other major European cities because of a dearth of similar studies, except for London where research conducted by C. Hamnett suggests similar results. It will be seen, however, that the few hypotheses to be advanced by way of illustration on how these situations come about indicate a not dissimilar pattern developing in large cities subject to the same level of economic activity.

Are social divisions in fact more acute in very large cities? This is a largely accepted view, which contrasts divisions in the big city with the more compatible, tightly-knit structure of its smaller counterparts. The question is not an easy one to answer, because methods of statistical analysis of urban social divisions are not readily adaptable for comparing one city with another. In the case of France, however, there are some pointers in N. Tabard's findings which cover the whole of the country. In the first instance, these show the socio-economic status of *communes* tending on average to rise with their size (1993: 15). For urban agglomerations, Paris stands out very clearly from other cities because of its very high average socio-economic status; 'almost three households out of four (72 per cent) from areas within the top 10 per cent of the socio-spatial ranking live in the Ile-de-France' (*ibid.*: 11). If one measures the internal differences in terms of status between the most well-to-do and the poorest areas in each urban area, the Paris region does not systematically show a greater difference;[9] it is the same or higher in, for instance, Marseilles, Lyons or Strasburg, and approximately the same for Toulouse or Nantes; it is, on the other hand, appreciably lower in the cases of Rennes, Amiens or Clermont-Ferrand. There would seem then to be no hard and fast case for concluding[10] that the larger the city the greater the division, nor that small or medium-sized cities are systematically more homogeneous. What characterises the Paris region is the strong presence of higher social categories, hence their visibility in the urban structure, whereas in smaller cities, well-to-do districts which produce the contrast upwards involve a relatively far smaller number of people. The comment should be made in passing that medium-sized cities appear in the light of these findings to present fairly marked differences from each other as regards internal social contrasts.

The other major dimension of social division in large cities is that of racial segregation. Whereas this has received very particular attention in the United States, research has been far less systematic in Europe, and especially in France, where statistical data are known to be limited; with census records showing nationality and place of birth, foreigners can be researched, but immigrants less easily (birth abroad will indicate French or foreign nationality, but how are *pieds noirs* to be distinguished from naturalised Algerian immigrants, and how indeed are these distinctions to be made for the succeeding generation?). It is further known that there are political, if not constitutional, reasons for this limitation; the Republican model for integration of naturalised immigrants requires that they be subject to no official discrimination. On the other hand, American sociologists have a whole range of census-based racial data available to them, the most revealing item perhaps the one that corresponds to the racial characterisation of ancestors: one is supposed to choose where one belongs, and so explicitly recognise one's membership of a specific community in spite of the intervening generations. What is there to be said about someone of mixed ancestry who is required nevertheless to make a single choice?[11]

However, thanks to these self-declared categories (they appear to be generally accepted and are seldom criticised by those who make use of them), which confuse 'race', language (Hispanics), and nationality of birth, it is possible in the United States to assess the racial composition of neighbourhoods with quite a fine grain. Despite the limitations of the French data, the degree of variation here between French and American cities appears much greater than where social class is concerned. Racial concentration would seem to be a lot higher in the United States, frequently more than 90 per cent in the case of some black neighbourhoods. By comparison, in the Goutte d'Or district of the XVIII *arrondissement* in Paris, one of the most 'ethnic' neighbourhoods, the 1982 census showed that out of 28,777 inhabitants there were 12,862 foreigners (35 per cent) and 1,438 naturalised French (5 per cent) (Toubon and Messamah 1988: 150). Furthermore, such racial specialisation is very selective in the United States, where there is limited mixing between different racial groups, quite unlike what is observed to be the case in Paris. To take the example of the Goutte d'Or again, if North Africans (Algerians and Tunisians mainly, then Moroccans) formed the largest group in 1982, other Africans figured fairly highly and there was an appreciable number of Portuguese, Yugoslavs, Spaniards and Indochinese (Toubon and Messamah 1988).

On this point too, Sassen's hypothesis, where she assimilates to a large extent the growth of a new tertiary proletariat and immigrant labour, making this a significant feature of the large 'global' metropolis, seems excessive, especially since 47 per cent of foreigners in employment in the Paris region in 1990 were in manual labour and only 22 per cent in white-collar jobs (INSEE 1990 population census).

However, one may well wonder whether the current 'failures' of the

Republican model for integration, which is linked to high unemployment and the crisis in education and in workers' organisations, may not harden and heighten racial divisions. The hypothesis made by E. Todd (1994), who sees assimilation and the preservation of differences as the consequence of anthropological structures established over a long period, may make for optimism in the French context, but it is not entirely convincing, nor is it conclusive, and the model does not rule out the possibility of discrepancies.

On the question of ethnic divisions in cities, and if one accepts Todd's typology of integrationist or differentialist societies, it is likely that no common European model exists. But, even more than in the case of class divisions, there is a dearth of systematic observation to provide a comparison between overall structures and the way they develop. More than in the case of the poor and the working class, monographs exist on immigrant districts, where the approach to social structure and way of life tends to be anthropological, but it is difficult to apply their conclusions generally to the combined urban structure of large cities. To date, cogent results are insufficient to allow comparison between the ethnic divisions in major metropolises, and since the real differences underlying statistical categories are more significant than with social and professional variables, such comparison is much more difficult to carry out.

Besides, even in France there is a dearth of studies comparing cities from this standpoint. It is known that foreigners are spread unevenly between regions as well as among social categories in city neighbourhoods (Desplanques and Tabard 1991); it is also known that their proportion increases with the size of conurbation (Chenu 1996: 228) and that their concentration in the Paris conurbation, which is far higher than in the others, representing 12.4 per cent of the overall figure in 1990, has increased during the last period between censuses, whilst it has dropped in the urban unit sectors sampled and in seven out of nine of the other conurbations of more than 400,000 (exceptions being Bordeaux, with a very slight increase, from 5.4 per cent to 5.5 per cent, and Nantes, slightly higher, from 2.4 per cent to 2.7 per cent, but where the level is well below the average (Chenu 1996: 229)).

But this greater visibility in the Paris region does not necessarily imply a more noticeable difference in distribution, which could only be measured by systematic comparative analysis. Neither can its social effects be interpreted with accuracy; a greater incidence of cohabitation with foreigners does not automatically produce an increase in interethnic tensions, contrary to pseudo-theories on 'tolerance levels'.

Processes of division

Analysing socio-spatial structures and their trend may be an important step towards understanding spatial social divisions in large cities, but it is not the whole picture; there needs to be an understanding of the processes which produce and transform these structures.

There would be few to disagree that segregation exists, but there are many different interpretations as to what causes it, and the debate is generally summarised as one between 'macrostructural' explanations and explanations which stress individual choice. The issues involved here often have more to do with politics than science: is this questionable feature of city living a structural effect of the capitalist social system (that needs to be reformed or replaced by a fairer system – the left-wing position); or is it rather the downside of a positive feature in our societies – individual freedom of choice in the market, whereby individual preferences, human nature being imperfect though rational, are held accountable, not the system (which is the best in the circumstances – the right-wing position)? The numerous analyses of the processes of segregation (which, be it noted, mostly deal with large cities, where segregation seems particularly to be an acknowledged but perhaps unaccepted fact) have identified three fields of investigation – public policies, the housing market and consumer practice.

Instances where segregation is directly attributable to a political and/or economic organisation with the specific objective of separating specific groups (caste systems, slave-owning societies, repressed religious minorities) have frequently occurred in the history of cities, but are infrequent in present-day developed capitalist societies, the violent interethnic conflicts in the former Yugoslavia being the closest in time. The apartheid policies in South Africa probably provide the most recent example, and without going very far back in the history of the United States one encounters explicit segregationist policies in the south.

More recently, most developed states have taken an unequivocal stand against segregation; hence the policies they pursue should in theory eliminate its likely causes. However, research on urban policies has frequently shown this not to be the case. While this is not the place to summarise findings, two principles would seem to emerge fairly clearly. First, the likelihood of policies inducing segregation can only be understood by taking proper account of complex interactions between various political actors and institutions, and among the different levels of government, acting in defence of interests and/or of various social objectives. Secondly, public policies invariably have consequences in – also complex – interactions with private actors, and in particular with the firms engaged in urban development. These two analytic principles are especially important in the case of large cities where the entanglement of political actors and levels of administration are more complex and where their relative unification in urban political institutions is more difficult to realise, i.e. in most European or American cities, where the issues they represent for the private producers in the city are more acute.

What then of the explanation offered in terms of the housing market, which is given significance by the classical analytic models of the social division of residential space inspired by the Chicago School? With the price of land and of housing being scaled in terms of desirability of location, choice of residence is a function of affordability and availability, hence the structuring of space reflects social division into classes. The cumulative inequality this gives rise to is acceptable insofar as, according to the American view of society, there is equality of opportunity, and social mobility. Inducing spatial mobility would make such inequalities theoretically temporary for any given individual or household.

In fact, explanation by way of the housing market immediately mobilises causal mechanisms which call into play other economic and social structures. Inequalities of income among households are largely determined by the structures and variations of the labour market, which leads on to the debate over the economic restructuring of large cities, which cannot be further developed here (cf. Pierre Veltz, chapter 1 of this volume).

The social definition of where it is agreeable or convenient to live depends, to some extent at least, on where firms and jobs are located. There are both positive and negative considerations, i.e. whether it is within easy reach of work, what drawbacks there are, and so on. Urban economic reorganisation brings about significant changes in the social use of space, depreciating or enhancing it (the inner city factories which close, so creating development opportunities). In greater Paris, there are clearly many contradictory features in the property market.

Competition between residential and economic use for desirable space will raise the price over a long period; if the reverse occurs, with pressure for lower prices because of a surplus of office accommodation or a depression brought on by earlier speculative investment, it is of short duration and limited in its effects. But desirability in itself has a scarcity value attracting prestige development for company offices, etc. (Pinçon and Pinçon-Charlot 1992). A high concentration of top management jobs also generates a corresponding demand for housing. The processes are complementary and, both by new building and refurbishment, bring about the social transformation of inner-city areas referred to above; the same pattern is discernible in up-market suburbs which, in various ways, reproduce a similar twofold desirability.

Conversely, the often antiquated state of the housing stock in Paris (half the residential apartments date from before 1915 and, according to the 1990 census, 17 per cent of housing is still without a bath or shower or an inside toilet) has given rise to a poor quality but relatively cheap housing market, which constitutes what specialists term 'de facto social housing'. This, coupled with the effects of the 1948 act controlling rents in old housing, has enabled a significant number of working-class households, whether in active employment or retired, to continue living in the city.

But the mechanisms mentioned above gradually erode this stock, some policies indeed aiding the process unintentionally. For example, attempts to situate major public amenities in eastern Paris, long advocated so as to remedy inadequate provision for these districts, has obdurately led to a rise in the symbolic and social value of the area, bringing in its wake piecemeal redevelopment which threatens to drive out those who live there; the siting of the Opéra Bastille and the several amenities of the new Parc de la Villette are cases in point.

The existence of a stock of social housing (HLM) in Paris and the inner suburbs is the other factor which enables a fairly substantial population on low incomes to live in the more central parts of the conurbation. But it is true that the spatial distribution of HLM is very uneven and tends to comply with the pattern of segregation. In 1990 12 per cent of Paris households occupied *HLM,* as compared with 21 per cent for the whole region and 32 per cent for the inner suburban department of Seine–Saint-Denis. And *HLM* inhabitants in Paris itself are less working-class, comprising 16 per cent middle management as against 12 per cent workers, compared with 6 per cent as against 31 per cent on average for the region and 4 per cent as against 35 per cent in Seine–Saint-Denis. Were these *HLM* not to exist, however, the 16 per cent of blue-collar, or 20 per cent of white-collar workers households in Paris itself would be hard put to pay the going market rents. Moreover, the large *HLM* stock in the inner suburban departments enables tenants, who for the most part are on low incomes, to live within easy reach of Paris, in many cases served by the Metro and fairly well provided with amenities and local public services (Pinçon, Preteceille and Rendu 1986).

Nevertheless, it is true that this state of affairs involves the existence and physical location of a stock of social, largely public, housing that is now around fifty years old; that, at the time when much of this housing was built, its location was more peripheral, hence less attractive, than it is now, the city having pushed out; and that the current tendency is gradually accentuating segregation: between 1982 and 1990, Paris recorded a rise of 7 per cent in the number of households living in *HLM,* comparative figures being 20 per cent for the inner ring and 25 per cent for the outer ring of suburbs.

If, therefore, the spatial distribution of *HLM* helps put a relative curb on social segregation at the present time in comparison with cities where the property market is wholly privatised, this is indeed a political effect, but one largely unforeseen or initially unintended, resulting in large part from a discrepancy between immobility in the spatial distribution of *HLM* stock and changes in the social utility of different areas of the conurbation. However, developments in the *HLM* housing stock raise questions as to the continued ability of this sector of the market to limit segregation. In addition to prevailing tendencies mentioned above, the dilapidated condition of many estates, the selective departure of the relatively more highly waged, and the management policy for allocating accommodation (cf. Oberti 1995) are giving rise to the increasing segregation of poorer families in more limited accommodation, with serious social consequences compounded by tenants' problems and the restricted conditions of living. Thus new forms of segregation appear in a number of suburbs while, at the opposite extreme, those with skilled or managerial jobs settle in new individual low-density housing within reach of new development areas where high skills are in demand.

Do the above remarks on the link between the housing market and segregation in greater Paris find an echo in other major metropolises in Europe? A start has been made with the comparative analysis of housing markets in European cities. We shall limit ourselves to commenting on some of the findings. In the case of London, the studies produced by Chris Hamnett (1984; 1987) seem to reveal certain likenesses in patterns of segregation and socio-spatial division. One proviso might be that there has been a radically new attitude to the policy of council housing over the last twenty years with regard both to new building and to privatising the existing stock and that this could well boost a trend towards relatively more segregation in the future. The higher segregation in Madrid referred to above (Leal 1990a; Preteceille 1993) can be related to the far more limited development of public-authority housing policy and resort on a large scale to home-ownership, this being a feature of the big cheap housing complexes on the edge of the city. In the case of Athens, T. Maloutas (1995) has shown that the lower incidence of segregation compared with other major metropolises is a consequence of the significant role of the traditional family structure in finding a home, the capitalist provision historically playing a reduced role.

These observations incline one to the opinion that the absence of a

'European model' of social division in major cities is probably connected with the diversity of systems for producing and distributing housing. A degree of convergence in this respect at the present time, itself produced by the process of economic globalisation in the cities, may well lead to similarities, but the underlying socio-spatial structures are fairly dissimilar and one should not play down their inertia effect.

Several times in this discussion we have been prompted to look at the features governing demand – at the 'preferences' of the consumers. It should be said at once that there are few, but well-established, cases where segregation is the indisputable result of individual choice. T. Schelling (1978) has proposed a theoretical model where segregation might result from the cumulative effects of individual preference for greater social or ethnic homogeneity in residential environment: effects unintended in themselves by each actor but which come about as a result of an unco-ordinated set of individual choices. One can only say that there would seem to be a limited number of concrete cases by which to verify the model (Clarke 1991); also that analysis needs to be taken beyond simply recording preferences, their genesis being a major field of sociological investigation. Individual racialism has a social history which produces it and reproduces it, as does the refusal or acceptance of cohabitation with other social categories. Then again, individual preferences count for less the further choice is restricted by social and economic structures.

The self-segregation of the upper classes, given concrete and symbolic expression in the districts of major cities favoured by the bourgeoisie, certainly provides the extreme case of household income not being a determining factor; the deliberate choice here to be with one's own kind is manifest (Pinçon and Pinçon-Charlot 1989). Even then it could be argued that the disinclination to keep private and business relations separate invests this social reserve from an economic viewpoint with considerable functional significance; or that self-enclosure in protected areas, guarded by private police and surrounded by a wire fence, as can be seen in suburban America or in Brazil (Davis 1990; Lopez 1996; Raffoul 1996; Caldeira 1996) is a response to the invasiveness of urban violence, itself a backlash to the repressiveness endemic in the economic and social violence of the system. Spatial polarisation such of an acute kind among the upper social categories has – thus far – not become a significant feature of European cities.

Working-class self-segregation and that of dominated ethnic groups are often represented symmetrically. The standard version is the culturalist–communitarian version; the deliberate grouping of like with like is claimed to have a function in tightening bonds, providing mutual help and maintaining the culture peculiar to the group. The most recent variant is that of the ethnic economic enclave (Portes and Bach 1985; Wilson and Portes 1980), though it can also apply to earlier periods and it is complementary to the previous variant: spatial regrouping enables an economic space dominated by the minority group to be constituted, in which wage and trade relations deriving from membership of the same community provide 'ethnic' firms with competitive and market advantages and wage-earners with access to a labour market and possibilities for upward mobility which would be more difficult for them in the mainstream economy.

The first variant is frequently mentioned in order to explain the black ghettos in the United States. But it is strongly contested, in the first place by the very people who attach value to the social and cultural effects of communitarian grouping. W. J. Wilson (1987), for instance, offers a partial explanation for the deteriorating situation within the black ghettos in terms of the departure of those who manage to achieve middle-class status: sufficient evidence that being in the ghetto is tantamount to a constraint, and that you get out as soon as you can. In the major European cities the dynamics of community solidarity are to be found wherever the densest concentration of immigrant groups is. But, particularly in the case of Paris, with the passage of time, the geographic mobility of immigrants, spreading them progressively throughout the city, is significant insofar as they integrate into French society, and the initial concentration dissolves much more quickly than in the United States, renewing itself chiefly with fresh supplies of immigrants. Leaving aside exceptional cases, like the recent case of the Mali immigrants in Montreuil whose desire was to continue living as a group and who declined the municipality's offer of rehousing, immigrants themselves do not wish to be further corralled in hostels or on estates, as is shown in different ways by M. Oberti's analysis (1995) on the 'residential strategies' of the working-class population in Nanterre and by a recent *CSA– Le Nouvel Observateur* survey. According to this survey, 88 per cent of the immigrants questioned said they would prefer to live 'in a district where there was a mixture of French people and foreigners', as compared with 6 per cent only 'in a district where there were mainly people from their own country'; similarly, regarding their children's education, 62 per cent prefer 'a French state school', as against 25 per cent 'a school teaching the customs or religion of their ancestors'.

The second variant presupposes that the members of the community dominate among the employers and employees of the 'economic enclave'. This is the case in the United States only with certain groups, among them the Miami Cubans, who provided A. Portes with his theoretic model. In the cases of New York and Los Angeles, J. Logan *et al.* (1995) have shown that the enclave model was applicable only to the Chinese and Japanese, Puerto Ricans, Mexicans and Blacks being very much underrepresented among heads of firms. In France where statistical analyses of the problem are very difficult for the reasons already given, it would, however, be possible to determine the level of wage-earners of different nationalities sector by sector, as one could also do in the case of employers. One would probably find significant linked concentrations of wage-earners and employers only in some sectors, such as building construction (Portuguese?), the retail food trade (Moroccans?), or in certain subsectors defined geographically or by the type of product – the clothing and food trade, for instance, with the 'Chinese' in the XIII Paris *arrondissement*. But such enclaves, which are far less important and far more the exception than in the United States, doubtless affect only a small proportion of immigrants.

These two types of explanation for self-segregation correspond to 'community' urban grouping, which is uncommon in major European cities. However, other types of individual 'preference' exist which may well produce relative segregation without there being marked political or economic pressure.

One would involve choice of lifestyle in different urban situations. With expenditure on housing matching income, a given household may decide between somewhere smaller to live but closer to the centre and to its cultural and other advantages, proximity of shops, etc., or somewhere more spacious and self-contained with what the suburbs offer – easier access to the countryside, to recreational facilities or to retail parks. Such considerations produce a considerable contrast in choices of location between households of the same social category depending on size of family, ages of children, etc. And similar income levels do not necessarily make for conformity; on the whole, those upper and middle categories in more 'intellectual' occupations tend to prefer living near the centre of town, while those in business or industry may prefer the suburbs.

Another influence on residential differentiation has to do with education. Higher or medium social categories who are more mindful of their status and their upward mobility, hence who will want their children to achieve, will be influenced by the quality of schooling available in the choice of where to live. This has become still more important with an increasingly competitive labour market and the emphasis given to high qualifications in securing stable and remunerative employment – all this against the background of the problems affecting state education.

A third factor – one frequently brought up in American literature – is local taxation. Public choice theory has made a particular point of this area affecting choice with the highly decentralised nature of financing and administration in respect of public education. Although in the Paris region there are considerable differences in fiscal pressure between one *commune* and another, largely because of differences in the tax base, and to a lesser extent because of municipal policy options (Preteceille 1993), there is little evidence of this being a factor in deciding where to live; though in view of the almost constant advance of local fiscal pressure, there is a possibility of its gradually becoming so.

The ability to exercise choice in the matter of where one lives is clearly a function of what one earns, earning power giving one access to a wider field; moreover, choice is affected by social situation. Hence there is a marked degree of interdependence between structural development and individual choice in producing socio-spatial divisions. Nevertheless, it would be wrong to underestimate individual choice even within lower social categories. For instance, the gradual movement of the population away from low-rental housing *(HLM)* dating from the early 1970s is in part explained by the desire for home-ownership on the part of households enjoying a regular and fairly high income in their category, and this as often as not implied a suburban move of some distance. Such households were helped by changes in housing policy, but they were not driven out of their *HLM*. And M. Oberti (1995) has shown that, at the present time, even those whose range of choice is severely limited (those seeking *HLM*) are able to express certain preferences and exercise choice for a particular estate, a particular block even, rather than another. However, correspondingly, it would be wrong to disregard structural evolution, as most econometric models do, and restrict oneself to the orthodox fantasy of free consumers in a pure market.

Social divisions, political fragmentation?

By way of conclusion, I shall look at the political consequences of social division in large cities. It will be a case chiefly of putting hypotheses and questions since, for all the criticism directed at segregation, there has been little systematic analysis of its consequences, especially in the context of politics where it has only been touched on in the course of studying voting patterns.

Hence returning to the question we posed at the outset, large metropolises more than any other type of city concentrate power, wealth, resources and abilities, as well as cultural life, and this gives them the potential to play a leading political role in the ongoing redefinition of the appropriate level for public action. By their size, by their complex and diversified economic structure, the quality of their infrastructure and their accessibility, by the specific advantages they offer, they represent major nodal points in economic globalisation and offer firms – multinationals and their associates in particular – strategic locations which cannot be readily matched elsewhere. In the intercity competitiveness set up by such firms, large cities seem to hold most of the assets, hence the greatest capacity for negotiation and for imposing public policies. Even so, these giants appear to be politically impotent, hampered by their size and complexity, where smaller cities present a strong identity and a strong image and are looked upon as being dynamic. Perhaps such paradoxical political weakness may result, as I suggested at the outset, from their social divisions which through fragmentation and conflict inhibit the emergence of a unifying political process.

A rapid appraisal of findings in these cities certainly reveals the extent of social division and the strength of the forces which reproduce and transform them. But it shows too that one cannot conclude that divisions are greater in the major cities than in the others: according to certain indicators some medium-sized cities display even greater divisions. If one could construct variables which made it possible to distinguish the relative ability of cities to become effective and unified political actors, these could be crossed with the various indicators enabling one to depict the extent of social divisions, and perhaps one could then distinguish an effective link between the two. In the present state of research, the point remains inconclusive.

Be it noted, however, that the results presented lead one to reject the pattern of marked social and spatial dualisation as being characteristic of very large cities. And the specificity of social divisions in the Paris metropolis has more to do with the sizeable pressure of higher social categories than with the lower ones. The leadership potential of such higher categories would further tend to make one reflect that unified political action ought to be more apparent.

If then the depth of social division is not an immediately useful variable, how are the specifically political features about which many observers are agreed to be explained? The size of cities has often been put forward as an explanatory factor, on the dual grounds that it sets local political institutions at a greater physical and symbolic distance and makes the problems of governing cities more complex and further removed from the everyday preoccupations of the population.

The first argument rests, in the French case, on a known outcome: the progression of the abstention rate in municipal elections with the size of *communes*. Furthermore, among *communes* of equal size, the rate of abstention is distinctly higher in the greater Paris region, which suggests that the size of the conurbation has a distinctive effect, and that it is not just an effect of municipal remoteness. It would be interesting to see whether the same effect is noticeable in the other major European cities where the division of urban districts is less acute.

On that basis one could point with approval to small or medium-sized towns – towns on a human scale, where the mayor is almost one of the family, and everyone knows each other – which are as different as possible from the loneliness and lack of identity in the great city. But, conversely, one could argue that the lack of interest shown in local politics springs from the antiquated territorial definition of municipal politics. Besides, the same differences in relation to size in the abstention rate in France do not apply to parliamentary or presidential elections; and the very large cities, in particular Paris, tend to be marked by intense political life. We shall return to this.

Inhabitants of big cities, apart from the oldest and the youngest and allowing for class differences, enjoy a spatial freedom far beyond their particular district in the many cross-town or out-of-town journeys they make. Their mobility, the places with which they identify, constantly outstrip their particular district and put them at odds with local politicians for whom the conurbation represents an assemblage of small cities. For the city-dweller, the relation of the parts to the whole, the way the city functions as a network, the interdependence of work and leisure, the criteria governing the quality of the environment, are all part of everyday life, and administrative divisions have little importance. For many local politicians the unit of the *commune* is all-important; inter-municipal cooperation represents a danger. The loss of interest in local politics is therefore the likely result of local administration being less and less adapted to new forms of city living.

The second argument, referred to above, is to do with the sheer complexity of government in large cities which makes it remote from the everyday preoccupations of its citizens and can be questioned for the same reasons. Problems of government may provide a partial explanation of the 'political weakness' of large cities, but for reasons other than citizens' indifference. Insofar as the large city constitutes a nodal point and a strategic point in economic functions and change and in current policies, the social and political issues which materialise there are probably more acute and open to more conflict than elsewhere. For the dominant class, the great city – the capital in the more centralised countries such as France, Greece or Britain, the larger cities in countries whose structure is polycentric, such as Germany, Italy or Spain – concentrates the functions of economic direction, political orientation and media domination; and because they have money to pay for the services they require in terms of urban environment and building construction, this represents a decisive asset for the capital growth within the city itself. It may be thought that, because of this and in spite of their electoral and political weight, it is less easy for these classes to establish political domination on the basis of consensus or compromise – to exercise hegemony in the Gramscian sense – because, on the one hand, the issues are too acute for them to make many concessions in regard to their interests and, on the other, those of the other classes are perhaps more sharply divergent. If one accepts that the political emergence of the city as actor presupposes in fact hegemonic governance, whatever social group occupies the leadership role, one may well consider that the intensity of contradictory interests related to the issues in large cities may produce a high potentiality for conflict and so make it difficult for the model to be realised.

Urban hegemony in large cities by the dominant class is likely to be particularly difficult and clearly leadership by others is equally unacceptable. And this might explain the political 'punishment' meted out to Paris by depriving it of an elected mayor from the time the Commune was crushed in 1871 until 1977 because of its revolutionary history; or else the dismantling of metropolitan government in London in 1986 with the abolition of the Labour-led Greater London Council.

Such potentiality for conflict possibly also stems from the firmer, more autonomous and more assured constitution of the other classes, the dominated classes especially, as political actors. The concentration and spatial distribution of the working-class population, first in the Paris *faubourgs*, then in the working-class suburbs (cf. for instance Brunet 1980; Fourcaut 1986; Girault 1977), certainly seem to have played an important part in the historic constitution of the labour movement. Nevertheless, the spatial pattern and concentration have only produced these effects in combination with a political culture and a body of practice exhibited in the city and inherited from its history. The political impact of the 'red belt', like the resistance it now displays, is explained not merely by the mechanical effect of the concentration of factories and working-class households but by a political history most often recalled across its most striking and symbolic moments – the French Revolution, 1848, the Paris Commune – but composed too of a sum of less spectacular gestures and exertions, wherein combinations play as big a role as divisions.

If Paris has thus been able to develop as a space for political protest, it is probably because the large city affords more space for political liberty: paradoxically, people are less under control, the degree of supervision is less, the constant intermixing of population makes it harder to assign a place to everyone and oblige them to keep to it. A further reason is that large cities, which are more open and more complex societies, continually offer the opportunity for confrontation and interaction with other social groups, stimulate mobility in all forms, indeed provide greater resources for social mobility – even if the risks are higher and defensive resources fewer. The growth of an ambitious working-class political identity probably owes a great deal – in spite of the mythology of the labour movement – to its associating with the intellectuals and the middle classes.

Rather than social division by itself, it seems to be a particular balance between division – hence a powerful enough concentration to permit autonomous political expression by the classes dominated – and combination – hence interaction, initiation, mobility, alliances – which characterises the political structuring of space in large European cities. Perhaps even the more open confrontation with the upper classes, whose existence and lifestyle, appropriation of space and command of resources have far greater visibility, contributes to its greater potentiality for political conflict.

One has to guard against a too facile association of this political conflictuality in large cities with paralysis, and political unification in cities of more moderate size with efficiency. The greater capacity for the autonomous expression of the interests of dominated classes may be considered to be a source of political innovation (consider the history of 'municipal socialism', which was particularly dynamic in the Paris region), and the potentiality for conflict to generate inventiveness and new solutions (supposing compromise is accepted as necessary, and not rejected as an obstacle to the dream of a complete upheaval). As against this, the sometimes suspect political unity of the medium-sized city can all too easily stifle the essential interests of those dominated and reproduce conservative political options, however these are dressed up as 'modernist'. Put differently, there will be more capacity for innovation in governance which is more disordered and open to conflict than in hegemonic governance which is more unified.

These conjectures, which draw their substance from the history of greater Paris, are, however, perhaps more open to question when one contemplates their future and that of the other major metropolises. Is it within the power of the middle classes, whose involvement in local political activity is greater, to give new life to working-class experience in Paris or in London or to effectuate future alliances between blue- and white-collar workers, and intermediate categories and a section of the higher categories?

Or else, with the higher categories being still more dominant and the working class much weakened numerically, economically and politically, with the past pattern of urban life changing and large cities becoming increasingly suburbanised, and with bipolarisation likely to become more acute and the social mix in danger of breaking up, it may be that large cities will undergo profound social change in a way that enhances governance but reduces the capacity for real social innovation.

G. Konrad (1996) drew attention to the historical continuity in the political personality of Budapest, to its cultural role as a great city, and to its lifestyle, in limiting the totalitarian tendencies of the socialist regime in Hungary. But perhaps out-and-out liberalism is poised to succeed there where totalitarianism failed. J.-L. Perrier (1997) observed that the literary cafés were now literary only in name and had become too expensive for those who had made them famous, and he quoted a Hungarian journalist: 'Before, poets and writers were sacrosanct. Now, the intelligentsia has no importance. All that is absolutely abhorrent and yet absolutely necessary.' To judge by the recent demonstrations against tough immigration laws in the past few years, Paris appears to be resisting this necessary trend. But observers noted that, for the first time, the social disturbances at the end of 1995 produced larger demonstrations in the provinces than in Paris.

Notes

[1] Other than that factorial analyses of heterogenous variables raise epistemological and theoretical difficulties, the results of such analyses, as well as the calibration of overall indices, rely strongly upon the definition, encoding and weighting of variables as well as in the size and shape of the design of spatial units in the study.

[2] Cf. for London, Congdon (1984; 1987); for Madrid, Leal (1990a; 1990b); for Paris, Bessy (1990), Preteceille (1993), Rhein (1994); for New York, Preteceille (1993); for Athens, Maloutas (1993, 1995); for Rio de Janeiro, Preteceille and Ribeiro (1999).

[3] Decline but not disappearance; there were still 1,184,007 industrial workers in the Paris region according to the 1990 census, i.e. 22 per cent of the total labour force – as many as in intermediate categories and more than in managerial or higher professional categories, with only white-collar workers exceeding them (1,586,977).

[4] Type I in the typology drawn up by Congdon, cf. Preteceille (1993: 70–1).

[5] I.e. the social grade corresponding to the first factor in Tabard's analysis (1993: 12, figure IIA).

[6] *Ibid.*: 13, figure IIB.

[7] *Ibid.*: 17, figure III.

[8] *Ibid.*

[9] I.e. the main ranking axis (figures IIA and B in Tabard 1993: 12–13).

[10] Other types of measurement might well produce different results, e.g. indices of segregation in each city.
[11] On ethnic categorisations in censuses and their links with differently originating forms of political management, see Simon (1997).

4 Social structures in medium-sized cities compared

Marco Oberti

Medium-sized cities, while they share much in common with larger ones, display specific characteristics as local societies. Even if the trend today is for markets and networks to expand on a national and increasingly a European scale, the distinguishing feature of medium-sized cities is still to be strongly rooted in a circumscribed and closely defined territory and so depend chiefly on local and regional networks. That is not to say that they are not attached to national or European groups, interests, markets and networks, but that their structure as local societies (with their political and economic actors, their history and traditions, their identity, their chief manifestations, their institutions, their urban fabric, etc.) remains a fundamental element of territory-based societal coherence – something that is increasingly elusive in large cities. These, by their sheer complexity, their internal diversity, their place in the global economy, the area they cover and their compartmentalisation, reveal themselves as very different urban societies in terms of overall coherence, visibility of social relations and social issues. Population settlement underlines the degree of contrast between the metropolis with its inflow of different cultures and races and its intermixing, and medium-sized cities which remain more homogeneous and more dependent on their surrounding regions. Similarly, the link that major cities establish with the global economy (financial markets, for instance) brings into being a whole range of activities themselves engendering professional occupations, new forms of labour organisation and new lifestyles which are felt less keenly in smaller urban communities. Such divergence of origin has its effect on identities, especially in relation to locality, whereas the notion of belonging to a particular place or region remains a factor of identification in smaller cities, especially when situated in regions with a marked identity. This is not to imply that those who live in big cities have lost all sense of locality, but simply that such a sense relates to a locality situated elsewhere, from where they originate. As a result, the sense of local identity is either reconstructed elsewhere and on other foundations or it is disclaimed.

However, recognising what is specific to medium-sized cities in relation to metropolises does not mean that they represent one and the same type of local society. Making use of a model drawn from a typology established by Max Weber (1921), and taken up and developed recently by Bagnasco and Negri (1994), I shall endeavour to draw a comparison between Britain, France and Italy in terms of the local societies attaching to medium-sized cities there and the link to be established between their social structures and the status and mobilisation of social groups. With their economic, political and cultural differences (Mendras and Schnapper 1990; Crouch and Streeck 1996), these three countries provide evidence that processes which are seemingly identical do not invariably produce the same effects.

Social structures and types of city

The typology proposed by Weber distinguishes the producers' from the consumers' city on the basis of the main classes of consumers and the nature of their income. In the city of consumers, those with private means or civil servants are the main economic agents. Their purchasing power largely conditions the activity of the local tradespeople or craftsmen. The city of producers on the other hand depends on the presence of factories and businesses. The entrepreneurs and, more particularly, the workforce as a whole represent the main consumers.

In the city of consumers, the presence of social strata which are 'not directly productive' but markedly consumerist influences the development of the city and its activities. In the city of producers, the city is structured by productive activities, hence consumer activities have less importance.

Thus, the definition and rationale of local social mechanisms draw from different principles – consumption and social practices away from work on the one hand, work and social classes on the other. In some instances the two principles may be superimposed; in others, with the middle classes for example, it is right that they should be distinct the better to identify social groups at the local level.

Industrialised cities with large-scale industry

Weber (1921: 22) presents the city of producers as the archetype of the modern city. He refers to cities which are already developing and prospering on the basis of industrial activity. Throughout the nineteenth century numerous cities, chiefly in Britain but also in France and Germany, developed out of industrialisation based upon large production units. A central element present was the emergence of a working class wholly dependent on that industry and its organisation.

Medium-sized cities whose development was bound up with this earlier form of industry today still bear the marks of the past in their social structures. Fordism in its purest form led to marked social polarisation. A working class was clearly distinct from a capitalist bourgeoisie. These two major classes were essentially contrasted in their ways of life, forms of consumption, places and types of residence and sociability, family structures, educational models and political attitudes. Corralled in their own districts in the centre or on the edge of town depending on the country, the workers were socially integrated through their labour, the modest consumption their wages allowed them, and strong representation across a powerful political and trade union structure. So long as large-scale industry thrived and guaranteed high levels of employment, the local working-class representatives more often than not confined themselves to defending the interests of the workers. Even if they referred to principles which went beyond the locality (capitalism and the class struggle), their political legitimacy and leadership were also based on their local presence. This probably constituted a basic difference from bigger cities where a large section of the working class came from other regions and other countries and where integration through labour was strong to the extent that local identity and local rootedness were weak.

In medium-sized cities where recruitment and social networks were more local and more linked to the surrounding region, local identity never entirely disappeared behind class identity: at least this was far less the case than in the big urban concentrations. In large cities, even after one generation, a worker's sense of local roots and the appearance of a strong territorial identity were always less significant than in smaller cities, which were able to integrate workers from outside by means of a gradual identification with their city or the region where they had found work. This difference in the way a worker related to a locality was still more marked when a low level of industrialisation led to essentially local recruitment as was the case in a number of medium-sized cities in France, or again when the form of industrialisation (built up on small firms, for instance) favoured local workers with a specific know-how, as in the areas of the industrial districts in Italy.

With the collapse of large-scale industry, these cities – and more particularly the ones most profoundly structured by industrialisation – would bear the full impact of the urban crisis, having to confront the problems of working-class representation, identification and culture, long-term unemployment, and at times the settling in of a large immigrant population and the rehabilitation of poorer residential districts which had deteriorated. The break-up of the social environment of large-scale industry led to a loss of integration in the poorer districts, initially in the civic and political sphere, with the crisis facing militantism and mass parties; then socially with the side-lining of the older generation of workers from the labour market and the insecurity facing the younger and less well qualified.

Medium-sized cities with a dispersed economy

Economic development based on large-scale industry continued to exert a considerable influence on the model of urban growth until recession overcame it and enabled new socio-economic models to emerge, frequently in regions which were already characterised by a small-enterprise culture or by traditions of independent labour. At the end of the 1960s, cities of producers based on the economic activity of small and medium-sized firms made an appearance in parts of Western Europe. In spite of their being economically both dynamic and innovative they never spawned a conurbation or local societies structured by Fordism and its social class system (Bagnasco 1988).

The 'Third Italy' embodies this productive model which is an alternative to Fordism (Becattini 1987; Fuà and Zacchia 1983). The return and the success of the small flexible firm are explained in part by a more differentiated and fragmented demand, a technology adapted to scaled-down production and improved instruments of communication. The central and north-eastern regions of Italy, where Fordism has had little impact, bear the hallmark of strong business, craft and small-firm traditions, while at the same time they present a rich array of medium-sized cities and banks which enabled urban functions to be dispersed (Bagnasco and Trigilia 1984; 1985).

Industrial districts specialising in one or more types of production formed in these cities which had the attribute of involving the local society, both labour and management, economically, culturally and politically. With shared values and shared expectations of success around the model of the small entrepreneur and the independent worker, mobilisation for economic development was intense. Furthermore, pronounced social mobility and fast and unfettered local relationships, both clearly reflected within firms, prevent the polarising of class structure and favour negotiation where there is an interplay of interests.

Social structure of this type differs greatly from that in Fordist society (Bagnasco 1986). In Italy, political subcultures joined in upholding conspicuous local identities, despite a clear perception of divergence in class interests. Such interests, whether in the more Catholic north-east or in the 'red' north–central regions. were redefined for the benefit of the locality and were perceived as being distinct from those of the industrial triangle of large-scale firms or from those of the underdeveloped south. The close relationships between workers and small entrepreneurs urged local governments to establish a balance between economic growth and the comprehensive transformation of local society. And the point needs to be made that such equilibrium was easier to establish because economic actors belonged to one and the same locality, with which they strongly identified and where intercourse was little affected by class considerations. These various elements combined to shape what Carlo Trigilia (1985) has called neolocalism. When the crisis occurred in industrial cities and threatened their social fabric, these smaller cities drew benefit from their social cohesion, so enabling the whole community to enjoy the advantages of economic development.

Both types of producers' cities then, whether dependent on large-scale industry or on a dispersed economy, have acquired their structure in depth on the basis of productive activity, even though the processes and the effects on locality are far from identical.

Cities which 'modernise'

Other medium-sized cities owe their development principally to services and administration. To qualify such cities as 'modernising' is in no sense to contrast them with 'productive' cities, thereby implying that these are obsolete. Far from being so, the cities in regions where there are small and medium-sized firms show a marked capacity to innovate and adapt to new economic conditions. Even cities where there is large-scale industry have brought about their transformation through modernisation, but they have done so in conditions which were far more unfavourable than applied in the cities I propose to call 'modernising' in order to underline the socially and politically dynamic character of an innovation which affects many aspects of local life.

The cities of the *ancien régime* in France were not involved in the process of industrialisation and subsequently prospered because they had a large class of consumers who were dependent on the state or on the services sector in general. In some of these cities, services accompanied the development of high-tech industries, which in themselves, in their form of organisation and their social characteristics, are very different from large-scale industry. Research, higher education and training play a decisive role both as regards innovation and the better exploitation of resources in production and as a reservoir of employment and a magnet for the better educated and more prosperous. Public and private services in firms therefore constitute essential assets in economic development.

In these cities, the principal actor is not the working class but a local bourgeoisie and more particularly a huge waged middle class which has played a significant role in the direction taken by urban development, not least by their aspirations, way of life and ground-breaking attributes. The decisive element is provided by local social groups who are defined by their social attitudes and by their place in the locality and not merely by their pattern of consumption or their professions. Having as they do an increasing amount of leisure, they devote much of themselves and their time to the life of the locality and have a social demeanour in line with their notion of citizenship and welfare (Bidou 1983; Vergès 1983; OCS 1987; Mendras 1994).

The weakness of the working class enables the waged middle classes in the public and private sectors to exercise an effective cultural and political leadership in organising and administering local life. The social structure is not polarised as in the Fordist city. Class distinctions are not something one is unaware of, but they are less clear-cut and stem chiefly from patterns of consumption, lifestyles and social attitudes.

Urban segregation does not take the form of so clear a division between solidly working-class districts with municipal housing in an often run-down condition and residential urban areas occupied by the bourgeoisie. The shaping of lifestyles and urban structures themselves do not simply reflect social class.

Because they are less dependent on large-scale industry, these towns have been less affected by the recession. The educational and cultural resources of the middle classes have proved to be decisively flexible and adaptable to the new requirements of the labour market. The sectors of production which underwent reorganisation, thereby endangering the livelihood of large sections of the working class, were anyhow of small importance, and the services sectors which subsequently developed were already well established.

Cities which are 'dependent and in receipt of aid'

In other contexts where a low level of local economic activity and insufficient resources are unable independently to promote economic development so as to create wealth and jobs, cities owe their development to firm intervention by government with little impact on the productive sectors. So it is with a number of southern Italian cities which are totally dependent on central government and its redistributive policies. Here is another type of consumers' city containing a huge public-sector middle class (white-collar and management), whose consumption drives an economy with its core elsewhere. Such consumption fuelled from the salaries of government officials and welfare benefits enables a whole range of independent activities – in trade and personal services – to be maintained. Such cities never take off economically. Their political dependence on redistributive policies binds them to central government which often exploits traditional cultural features (kinship networks, protection tie-ups, membership of a local community, etc.) to reinforce a loose form of clientelism. Here we are far away from the innovative 'modernising' cities mentioned above, in which a highly qualified bourgeoisie from the public as well as the private sector provides the essential economic drive, not simply as consumers but because their assets are invested in the community. These are cities 'in care', dependent on discretionary political management, and as such incapable of independent economic development.

In actual fact, almost all cities contain mixed features. Cities with large-scale industry or which are economically dispersed cannot be baldly assumed to be organised entirely on the basis of production or of social class, any more than cities of consumers can be seen as depending exclusively on services and middle-class lifestyles and involvement in the local society. Different types of activity – in industry, services, commerce, finance, etc. – co-exist in all cities and the effects of consumption and of the link with services, both public and private, do not only count in cities dominated by tertiary activities. Taking the model as a basis, I now propose to specify types of urban society which, though they may have features in common, can be distinguished both within one country and between one country and another.

The model in the context of three countries

In Britain, Italy and France, the different types of city do not have the same significance and sometimes take on different forms. All types are not to be found in all countries, in one country certain types may dominate, and one and the same type may reflect different social realities.

Britain

The early occurrence and the sheer scale of industrialisation in Britain produced, more than in other countries, large and medium-sized industrial cities which come closest to our model of the city of large-scale producers. Alongside the major cities of the industrial revolution

(London, Birmingham, Manchester, Liverpool, Glasgow, Cardiff), a number of other cities followed this model. Leeds, Sheffield, Newcastle-on-Tyne, Bradford, Coventry and Wolverhampton, among others, after enjoying the prosperity which industry brought, had to suffer the full impact of deindustrialisation and experience the acute effects of urban crisis. If Britain defines itself more than other European countries as a society based on class, it is precisely because these industrial cities developed with the marked characteristics of a strong traditional working-class presence. The upper middle classes, the bourgeoisie, belonged to a different social world, with their homes in the residential suburbs or, for the richer among them, in small towns in the surrounding countryside. But living and working in the city as they did, the working class set the 'tone', politically, culturally and socially. As a general rule, in the medium-sized industrial cities, middle classes from the private sector were little represented, unlike those from the public sector who were culturally less distant from the working class. Moreover medium-grade jobs in the public or municipal services were generally held by people who came from a fairly modest background, particularly in social services, this being less evident for the same grade of job in the private sector. Thus, for these employees at least there was a degree of affinity with the working class, with whom anyway most of those working in social services were concerned more directly. This is not to say, however, that political opinions or those on local issues were shared.

The urban crisis and the industrial recession which struck these cities in the 1970s significantly reduced the room for manoeuvre open to local social groups and, firstly, to those with political responsibility, who were left with no other alternative than to give priority to those most affected by the recession. Local initiatives and measures taken focused on this, frequently to the detriment – in the early stages at least – of investment and reforms in the fields of education and culture and advanced services for industrial firms, all of which during the 1970s and 1980s were turned to the advantage of cities that had been spared the worst of the recession.

In the trough of the industrial recession in Britain, when the economic and social situation called for a strong political response, the political representatives of the working class championed a class position and programme with which the middle classes did not always agree, even those from the public sector who put forward other points of view while ostensibly defending working-class interests (Le Galès 1993).

From the 1960s onwards, reflationary measures benefited services, more especially in the private sector, and principally medium-sized southern cities: Gloucester, Winchester and Canterbury, for instance;

Swindon and Reading; new towns like Milton Keynes and Stevenage; newly urbanised towns like Peterborough; and cities with a long history like Norwich and Ipswich. Just when most of the northern cities saw their economy collapse, with firms and the middle and higher strata of management leaving and the threat of social disintegration made real by the loss of jobs for a large part of the workforce, southern cities were experiencing new prosperity, driven by services and new technology and attracting new higher-waged employment. Nor were they faced with the political and social cost of the damage inflicted by the recession, but rather were able, with a head start and without constraint, to develop policies aimed directly at the middle classes and growth sectors, thus providing themselves with the infrastructure and resources to reach for the top of the market. In these cities, a degree of antagonism developed within the middle classes between those in the private and those in the public sector over what the development model for the city should be, those in the private sector, less frequently of working-class origin, tending to adopt the outlook common to small businesses, and those in the public sector tending to identify with white-collar workers, and hence more receptive to social and welfare issues.

Just as in France, those in the middle classes linked to the private sector were more favourable to economic development, their counterparts more naturally in tune with social or cultural issues. Nevertheless, in both countries, consumer patterns and social attitudes were largely instrumental in influencing urban policies. Services related to leisure activities and higher education were well developed, while many other areas – housing, transport, quality of life – tended to reflect the posture of groups whose position in the city was determined less by economic considerations than by general involvement in the life of the community. A comparison between Coventry and Norwich gives a good idea of the two sides of medium-sized cities in Britain. Coventry, which has a long industrial tradition (and was the chief seat of the motor industry) was recently forced to work out a development plan so as both to include and exclude its working class, against the background of a multitude of job losses and industrial dereliction. Norwich, which early on opted for encouraging services and new technology, had no problems of this order to confront and has been able to attract middle-grade employment.

France

In France also, middle-rank cities affected by the run-down of industry had as a priority to face the consequences of job losses on working people and their living conditions. With central government playing a more robust interventionist role, the damage was less than in industrial parts of Britain. Nevertheless, just as in Britain, urgent social and welfare measures somewhat obstructed the more enterprising political, economic and cultural action needed to attract the activities and the grade of manpower previously underrepresented. St Etienne and Le Creusot are cases in point, along with comparable cities in the north. Outlying industrial suburbs too – Mantes and Dreux to the west of Paris, for instance, which grew in the wake of new industrial locations – were affected by economic restructuring.

St Etienne, with its strong tradition in the steel and metallurgical industries (arms manufacture, in particular), was badly hit by the recession, as was its workforce, much of it foreign, accommodated mainly in outlying parts of the city. Faced with the urgent need to devise a rescue plan, the city endeavoured to change its run-down working-class image and modernise its economy, but the policy has been somewhat self-defeating and has merely aggravated the social and spatial fracture.

What seems to be most specific to France is a particular type of consumers' city: a city of medium size, less affected by the recession, where a mainly middle-class public-sector workforce predominates, whose flavour, lifestyle and municipal involvement the city clearly reflects. Rennes well represents this type in its social make-up, where both the working class and the *grande bourgeoisie* are relatively unimportant. And although industrial development in Rennes exemplifies the regionalisation policies for industry of the 1950s and 1960s, the city is primarily administrative and service-orientated. Being less under the shadow of the recession, the predominantly middle-class population, which is well represented in the municipality, has done much to promote local development and a sense of community in cultural and leisure activities and in environmental life, at the same time setting out to attract high-tech industries (Le Galès 1993). Such unselfconscious image-making appears to have done much to boost the social and economic well-being of this type of medium-sized city.

Larger cities are tied up with economic and political considerations of another order which deny this role to the middle strata; and medium-sized industrial cities, where class differences are too marked or where the working class has been much affected by the recession have not only shown themselves to be less attractive to these middle strata but also to be in need of further political and cultural investment. The type of medium-sized tertiary and public-sector-based city illustrated above is well represented in France. They tend to be long-established regional capitals and to be provided with a university and its resources.

Italy

The case of Italy is very different from that of France or Britain. Firstly, large-scale industry has been chiefly restricted to the Turin–Milan–Genoa triangle. Medium-sized cities have been important in developing an economic model for industry based on small and medium-sized businesses (the dispersed economy), but it is very much the exception to find they have large-scale industry. In general, cities of this type, though they represent producers' cities, have escaped the recession affecting large-scale industry, and to a certain extent have even benefited from it and have managed to achieve a high level of social integration. As urban societies, they are strongly rooted in their locality and motivated by economic activity to which total commitment is given. Even if interaction at a local level is class-structured (employers and workers), a highly integrated workforce which has been little affected by the recession and the absence of 'disaffiliated' persons (Castel 1995) provides a guarantee of social cohesion, which is further cemented by a strong sense of local identity. Certainly today it is the one type of producers' city to have escaped the urban crisis that most industrial cities have been unable to avoid.

On the other hand, there are few medium-sized cities corresponding to the French consumers' city model with its 'modernising' strain, since compared with other European countries, the social structure in Italy contains a high proportion of self-employed (craftsmen, small businessmen, professional people and so on) and relatively fewer waged middle strata, who thus have less impact on lifestyle than they do in France. Be that as it may, depending on the importance of services in producers' cities, on whether they are public or private and hence on the prominence of this waged class, the social profile of such cities is variable and may sometimes draw close to the consumers' city model to be found in Britain and still more clearly in France.

Thus in economically dispersed regions one finds cities which have accumulated significant administrative and tertiary functions, as has Perugia, capital of Umbria. A university, teaching hospital, research institutions and public and private administration draw the kind of social groups whose demands on the locality are new, or at least different from the more traditional ones voiced by those who run small businesses. The economic argument and outlook gives way to one of social differentiation based upon consumer activities, including the initiatives shown by the highly educated public-sector middle class. Areas which might appear marginal in relation to small business activities, such as culture, the quality of life and the environment, the use of leisure, etc., and, added to this, a more outspoken strategic view as to how these concerns can be given expression in local development, become gradually incorporated into political action. The life of the community is thus enriched very much to the benefit of the large student population, and this is clearly visible in the development and social organisation of the city.

In striking contrast is Bergamo, in the east of Lombardy, in a region of small and medium-sized firms, but where there are also important tertiary activities. A highly qualified waged middle class is to be found there, but seems for reasons that have to do with local culture (one being the view held about the entrepreneurial role) to be somewhat removed from the middle-class model. It may well be that in Bergamo attitudes among the waged middle class are different from the traditional small employers but they reflect close attachment to notions of work, saving, investment and success which inform the small business world. Bergamo as a local society is still largely geared to economic activity, and within it the social, political and cultural autonomy of the waged middle classes seems less pronounced and less decisive a force for change in the city. These features, added to others such as a marked Catholicism, a regional dialect, a concern for family values, etc., consolidate a local culture which puts secondary value upon free-time activities. And, unlike Perugia which nurtures its role as a cultural capital, with justly celebrated festivals, a renowned university, a variety of exhibitions in the arts and so on, Bergamo promotes itself above all as a city of production, hosting agricultural and industrial fairs at a national and international level, in spite of the fact that its artistic heritage is far from negligible.

The type of producers' city with a social and urban structure which is largely influenced by consumer classes, here shown by the case of Perugia, illustrates a more global tendency for change in Italian society, but which has taken different forms from tendencies observed in France or in Britain. The very particular character of urban societies in the 'Third Italy' seems to merit more exhaustive comparative study.

Central government in Italy is involved on a massive scale with the medium-sized cities of the *mezzogiorno* so as to make up for their economic torpor. For complex reasons (Trigilia 1992; Bevilacqua 1993), state intervention has rarely produced the effects expected in market economy terms. In fact, it has frequently brought about a situation of political dependence, with government limiting itself to pouring out large sums of public money in the form of welfare assistance or new jobs in administration, or modifying legislation and taxation to suit small businesses or the professions. Political intermediaries would rely on kinship networks and contacts of all kinds, sometimes including the Mafia, to put together what can only be called a clientelist political administration. A whole stratum of government employees and many local authorities thus prospered artificially, in the sense of having no real links with the local economy, since much of their consumption related to products manufactured in the north. And the economy depended entirely on the state in order to function. Of course, not all *mezzogiorno* cities in this category conformed to this state of affairs, and economic and social development in the south is itself geographically diversified (Trigilia 1992), particular areas being economically more dynamic and far less dependent on government, on political clientelism and on a criminal economy.

Those most at risk among the population were also in thrall to political clientelism through a discretionary but generous system of awarding benefits, allowances and pensions. The picture is very different from the 'modernising' cities of consumers where interdependence between public and private services and an innovative productive sector have a dynamic effect on local development. Such towns might rather be depicted as being publicly disorganised and lacking forceful private initiative; certainly they afford striking contrasts: unregulated town planning, districts ill provided for and with few amenities, where side by side one can find well-appointed residential areas, a deplorable public transport system, inefficient and corrupt public services, a well-established criminal economy and so on – evidence enough that the presence of one class of consumers is inadequate to represent the cities of the *mezzogiorno* and in particular to compare them with urban contexts elsewhere in Europe.

Conclusion

This brief summary of the social structures of medium-sized cities in three European countries allows the following conclusions to be drawn.

The models for economic development in each of the three countries (where the rhythm and degree of industrialisation and the level of state intervention constitute two major dimensions) have had a considerable influence on urban development. Medium-sized cities fall into different types and are differently represented from one country to another.

Those cities least marked by the industrial recession have been significantly influenced by the groundbreaking lifestyle of the waged middle classes. National differences here count for something because it is not everywhere and invariably the same strata of the middle class who are most prominent. In France, the importance of the public sector has promoted a large waged middle class, whose expectations and values have been very influential in medium-sized cities, in their style of administration and their development, both economically and as a community. In Britain, with the retreat of the state, a similar role has in the main fallen to the private-sector middle class in the southern half of the country. This course of events is still far from common in Italy in spite of there being the strong presence in some cities of a waged middle class which is more independent of the culture traditional to the dispersed economy.

On the other hand, medium-sized cities affected by the crisis in large-scale industry have left little space in the locality for the development and self-assertion of waged middle classes who, while being well represented in local government, have been faced as a first consideration with the social problems consequent on the run-down of large-scale industry. The past pattern of the industrial working class has counted for much, the type of city suffering from the double blow of industrial recession and the disengagement of government (hence of middle classes) being particularly characteristic of Britain.

Italy is distinct from Britain and France in that the waged middle classes play a much less dominant role in the social structure of medium-sized cities, and in particular those of the industrial districts which are so much a feature of central and north-eastern Italy. The waged middle classes are as much if not more consumerist than in other European countries, but given that their political representation and action are relatively independent of other groups in the locality, they do not find it easy to constitute a social group. In comparison with Britain and, more particularly, with France they appear to be less committed to the idea and furtherance of culture, of a certain lifestyle, and of the city's image. This is not to imply, however, that such Italian cities lack a cultural life or value the quality of life less; indeed more than elsewhere in Europe they comprise genuine urban communities. Social regulation within the locality is doubtless less institutionalised, but its informal, discursive nature, where values are shared and there is a strong sense of belonging, sustains a degree of social cohesion which has diminished in a great many cities in Europe.

The medium-sized cities of the *mezzogiorno* represent a particular type of urban society, unlike any in the other two countries. Lacking focus on economic initiative, unco-ordinated and clientelist, government intervention, linked as it is to traditional structures and reactivated by a criminal economy, has not been conducive to the emergence of a waged middle class prepared, as in France or Britain, to take on a role in urban development.

5 Different cities in different welfare states

Juhani Lehto

There is a tendency in the recent social science literature on cities to assume that global economic restructuring creates similar urban development in all industrialised or post-industrial countries. There are different trajectories for different categories of cities such as 'global cities', 'regional centres', or 'dying former industrial centres', which are explained by their different positions in the global economy. At the same time the similarities between the cities of different countries and the differences between the similar cities of different countries have received less attention.

It is the basic assumption in this chapter that national institutions, particularly the welfare state, have shaped and will continue to shape the development of cities. The trajectories of the different categories of cities are assumed to be different within different welfare states. Thus, there is not only international convergence in urban development processes, but there is also significant divergence across different welfare states. Three major themes related to the basic assumption about the impact of the welfare state on the city are discussed. The first is related to the social and spatial history of cities and the impact of the welfare state on the process of urbanisation. It is argued that cities that developed before the expansion of the welfare state are different from cities that developed parallel to the development of the welfare state. The second theme is related to social and political structures within the cities and to the impact of the welfare state on the shaping of interests, political actors and partnerships in cities. It is argued that the welfare state, as a major institution in the production of services and the shaping of opportunities for consumption, should be taken into account when studying the shaping of actors in cities. The third theme is related to the changing relationship between national states and cities. Building the welfare state has been a significant part of the building of the national states in Europe. It is argued that differences in welfare states may have a significant impact on the process of renegotiating the role of cities in the national and international arena.

Comparative welfare state research has identified different welfare state 'regimes' (Esping-Andersen 1996). Quite often, a particular Scandinavian welfare state 'model' is presented. The Scandinavian countries have had high social expenditure, effective redistribution of income with small income differentials and low poverty rates, large public employment in education, health, social and other services and broad consensus concerning their social policies. It might be expected that the impact of the welfare state on cities would be most visible in Scandinavia. This, in addition to the fact that the author is from Finland, might be given as an explanation for the approach in this chapter, which is mainly based on discussion about cities in Scandinavian welfare states, with only some comparisons to cities in continental Europe, the UK and the USA.

Urbanisation before and during welfare state development

According to Therborn (1995: 68–9), Europe is the only part of the world that has developed from an agrarian society to the present service society via a phase dominated by industrial employment. In other parts of the world there has never been a period of relative preponderance of industrial over agrarian and service employment. For instance, the USA, Japan, South Korea, Chile and Argentina moved directly from an agrarian to a service society.

The dominant industrial period lasted from 1821 to 1959 in the UK and also lasted a long time in Belgium, Switzerland and Germany. In other European countries it came much later and lasted for a much shorter time. Countries such as Denmark, Finland, Greece, Ireland, the Netherlands, Norway and Spain followed the direct road from an agrarian to a service society, according to Therborn's interpretation of OECD labour force statistics (see table 5.1).

Table 5.1 also indicates that the peak year of industrial employment for all countries other than the UK was after the second world war. It was not before but during the greatest expansion period of West European welfare states. It was also not before but during the change towards a service society in Western Europe. The change towards a European service society and the expansion of European welfare states are also closely interrelated. A significant proportion of the service jobs are provided in education, health and social services, largely funded from the welfare state budget.

Certainly countries other than the UK have old industrial centres, which were developed before the expansion of the welfare state and which have had a clear dominance of industrial employment. However,

Table 5.1. *Period of relative preponderance of industrial over agrarian and service employment and the year when industrial employment was at its peak in different countries (Therborn 1995: 69).*

Country	Industrial period	Peak year of industrialism
Austria	1951–66	1973
Belgium	1880–65	1947
Bulgaria	1965–	1987
Denmark	Never	1969
Finland	Never	1975
France	1954–9	1973
Germany (FRG)	1907–75	1970
Greece	Never	1980
Hungary	1963–88	1970
Ireland	Never	1974
Italy	1960–5	1971
Netherlands	Never	1965
Norway	Never	1971
Poland	1974–91	1980
Portugal	1982	1982
Romania	1976–	1980
Spain	Never	1975
Sweden	1940–59	1965
Switzerland	1888–1970	1963
UK	1821–1959	1911
Argentina	Never	1960
Japan	Never	1973
USA	Never	1967

most European urbanisation has occurred in a different context, parallel to a change towards a service society and the expansion of the welfare state.

It may be assumed that a long period of industrial dominance before the expansion of the service sector and the welfare state has had an impact on the spatial structure and the political culture of a city. It is probable that such a city has old industrial areas and old working-class housing very near the centre that give a particular significance to inner-city problems and development. It is also probable that there is or has been a strong impact of traditional working-class representation in city politics and development. And it is probable that deindustrialization has had a larger impact on the development of such a city.

Urban development parallel to the expansion of the service sector may be expected to lead to a city centre dominated by services and a dispersion of industries and housing to separate areas at varying distances from the centre. Also social problems may be scattered in different areas of the city, both in the centre and in some suburbs. The new urban middle classes are expected to have had a stronger influence in shaping the political traditions of the city, and deindustrialization may have a weaker direct impact on such a city. Social development is more strongly linked with socio-economic changes affecting the 'new middle-class way of life'.

The welfare state has a large spectrum of linkages to the development of cities. It has impact on the social divisions within cities by redistributing income and welfare. It offers jobs, particularly in education, health and social services. It may enable female participation in wage labour by offering social services that reduce the need for the unpaid care of children and the elderly. Income transfers, jobs and services also decrease the economic risks and increase the consumption potential of families. Thus, they provide opportunities for a 'middle-class pattern of consumption' including, for instance, home-ownership and a private car, which have impact both on the physical and on the social structure of cities. Sometimes this significant role of the welfare state has been forgotten in analyses which explain urban development almost as a deterministic reflection of the change from an industrial or Fordist economy to a post-industrial or post-Fordist economy.

Including the welfare state in the analysis also underlines the fact that different institutional contexts and politics may matter in urban development. Thus, urbanisation was different before and during the expansion of the welfare state. Some of the differences in urbanisation between industrial societies and service societies are also related to the different impact of the welfare state on these societies. It may also be assumed that different welfare state institutions and policies result in differences in urban development.

The institutional and policy differences between welfare states are the basis for a growing literature of comparative welfare state studies. The differences between the Scandinavian, continental European and Anglo-Saxon welfare state 'models' have received much attention. Subsequently, it has been shown that welfare state institutions and policies also differ within these three groups and that there are other additional 'models' such as the emerging East Asian welfare state in Japan, South Korea and Taiwan (Esping-Andersen 1996).

Different welfare state institutions and policies are related to different rates of poverty, differences in social class structure, mobility between social classes and differences in income distribution within the

Table 5.2. *Income differentials in some OECD countries measured by Gini index (the greater the index, the greater inequality with regard to available income per inhabitant), in 1986–7 (Atkinson et al. 1995).*

Country	Gini index
Finland	20.7
Sweden	22.0
Norway	23.4
Belgium	23.5
Germany (West)	25.0
Netherlands	26.8
France	29.6
United Kingdom	30.4
Italy	31.0
USA	34.1

Table 5.3. *Indicators of female participation in wage labour in some OECD countries in 1986–9 (Julkunen 1992).*

Country	Percentage of women in labour force	Percentage of female workforce working part-time	Percentage of mothers of 7/10-year-old children working outside home
Sweden	80	45	87
Denmark	77	42	79
Finland	73	11	79
Norway	71	45	69
United Kingdom	65	45	46
Portugal	60	7	62
France	56	23	56
Germany (West)	54	31	38
Netherlands	52	51	32
Belgium	52	24	54
Italy	44	10	42

population (see table 5.2). The differences in the extent of social care services and in the rate of female participation in wage labour are also quite significant (see table 5.3).

The US welfare state allows significantly greater income differentials and a higher poverty rate than the continental European or the Scandinavian welfare states. European public authorities also have wider powers to guide the housing market and local planning (Therborn 1995). These differences in welfare states explain at least part of the greater social and spatial divisions between the rich and the poor citizens in US cities compared with the European cities.

There are considerable differences between Scandinavian and central European welfare states, too. The Scandinavian welfare state has kept poverty rates and income differentials lower than in the rest of Europe. It has also enabled females to participate in wage labour and offered more jobs to women in the extensive public-service sector. Also the overall rate of participation in wage labour in the working-age population has been considerably higher and unemployment lower in Scandinavia, at least until the 1990s (Esping-Andersen 1996). Housing policies have also been more deliberately planned with the aim of preventing deep social divisions in cities (Harloe 1995; Lankinen 1994).

The golden age of the welfare state in Scandinavian cities

The most rapid urbanisation and the expansion of the welfare state were parallel phenomena in Scandinavia. This period started first in Sweden, which remained outside the second world war, and last in Finland, which joined the Scandinavian urbanisation process only in the 1960s and 1970s. As cities and the welfare state grew in parallel, the welfare state was not only a response to social problems related to industrialisation and urbanisation, it was also a significant factor in attracting people away from rural life into cities. A high proportion of urban growth was channelled to new suburbs and satellite towns.

The new suburban families had to invest in their home and a private car for moving between their work place and home. Such investments demanded the participation in wage labour of both men and women: two 'breadwinners' in an ordinary family. Thus, it was necessary to create jobs for women. And if women were to search for a job outside the home, the demand for social care services, such as day care for children and care for the elderly, had to be met by services provided outside the household and the family, and there also had to be social security in case of sickness, invalidity and unemployment. Otherwise, the investment in a flat or house, in a private car and in other aspects of the 'middle-class way of life' would have been too risky for large lower segments of the new urban population.

The expansion of the welfare state guaranteed the necessary stability of income. It provided most of the jobs for women and it provided the care services to enable women to participate in wage labour. While unemployment started to grow in many other West European countries from the mid 1970s, Denmark entered this phase only in the mid 1980s and Sweden and Finland only in the early 1990s. The main explanation for this difference seems to be that the stagnation in private-sector employment was compensated for by a continuous increase in public sector jobs in Scandinavia (Esping-Andersen 1996). The Scandinavian country with the latest expansion of the welfare state, Finland, has been the extreme case in this regard. Its total employment increased by 10 per cent, employment by the state by 31 per cent and by the municipalities by 193 per cent, between 1970 and 1991 (Kasvio 1994).

Although the development of the welfare state was a significant factor in the Scandinavian urbanisation, social policies were not labelled as 'urban policies'. On the contrary, a firmly stated goal of the Scandinavian welfare state has been to create equal opportunities for using welfare state benefits and services in terms of socio-economic groups, central and peripheral regions, urban and rural population. To label social policy as urban would have been against this universality principle. A compromise between 'urban' and 'rural' social policy has also been necessary because most Scandinavian governments, although led by social democratic parties, also needed support from agrarian bourgeois parties. Only the Swedish social democratic party has succeeded in gathering enough support to govern alone for long periods, but it has also needed support from rural Sweden. Thus the economic and employment role of the welfare state has been even greater in rural than in urban regions.

The 'golden age' of the Scandinavian welfare state meant that the situation in cities differed considerably from the situation in many other West European cities in the 1980s. The unemployment rate was low, and much lower in cities than in less populated areas. Income distribution was even and the proportion of the poor in the population was much smaller than in most other OECD countries. With regard to immigration there were considerable differences between the Scandinavian countries. Sweden had a considerable foreign-born population, which had mainly come from other European countries. The other extreme was Finland, a country of net emigration, mainly to Sweden, until the mid 1970s. In any case, the Scandinavian countries and their cities have been considered to be exceptionally homogenous, in socio-economic as well as in cultural and ethnic terms. Cities in Scandinavia are small in comparison with other countries, and even the metropolises are quite small (see table 5.4). Thus, there has been much less history of social segregation in the cities than in many other countries.

It has also been a stated goal of Scandinavian housing policy to prevent segregation by mixing social housing, co-operative and private rented housing and private home-ownership. A thorough study of the

Table 5.4. *Scandinavian cities with more than 150,000 inhabitants in 1992–3 (Nord 1994).*

City	Population	Percentage of the total population of the country
Copenhagen metropolitan area	1,343 000	26
Copenhagen	620,000	12
Århus	271,000	5
Odense	181,000	3.5
Helsinki metropolitan area	848,000	17
Helsinki	502,000	10
Espoo	179,000	3.5
Tampere	175,000	3.5
Turku	160,000	3.2
Vantaa	160,000	3.2
Reykjavik metropolitan area	151,000	58
Oslo metropolitan area	736,000	17
Oslo	474,000	11
Bergen	218,000	5
Stockholm metropolitan area	1,520,000	18
Stockholm	685,000	8
Gothenburg	434,000	5
Malmo	237,000	2.7
Uppsala	175,000	2.0

development of socio-economic conditions in different subdivisions of the Helsinki metropolitan area proved that the 1980s was a decade of decreasing differences and desegregation (Lankinen 1994). More pessimistic evaluations have been expressed about the development in Sweden and Denmark. However, even the more pessimistic evaluations identify the impact of Scandinavian anti-segregation policies (Öresjö 1995; Andersen and Munk 1995). Thus, urban segregation in Scandinavia has mostly been discussed from the perspective of 'how to prevent similar developments which have created greater problems in the UK, other European countries and the USA'.

An end of the golden age?

Scandinavia has not survived global economic restructuring and the ideological and political trends of the Western world without changes or problems. Denmark experienced a rapid increase in unemployment and strong efforts to cut social expenditure, in the mid 1980s. Finland and Sweden reached the same phase in 1991–2. Denmark has not succeeded in returning to the previous low level of unemployment, and high or much higher than previous levels of unemployment also seem to continue in Finland and Sweden. Even Norway is experiencing an increase in unemployment, although it has no public deficit problems, due to the country's vast oil revenue.

Although there has been much concern and discussion about the 'collapse of the Scandinavian welfare state' and 'the growing social divisions in Scandinavian societies', due to the increase in unemployment and cuts in welfare expenditure, the actual changes seem to be smaller than expected. For instance, inequality in income distribution and the poverty rate have not significantly increased, at least not yet. The safety net of the Scandinavian welfare state seems to be holding, even at a time of high unemployment (Kautto *et al.* 1999). However, it may be too early to say anything about the long-term impact of high unemployment in Scandinavian societies and cities.

Different welfare states – different actors in cities

A traditional sociological analysis of the interest groups and political actors in cities was based on concepts such as social classes or strata and their political or ideological programmes. Later, the position of citizens as consumers of commodities, space and environment has also received attention in the analysis of actors in cities.

The welfare state has a significant impact on the shaping of actors in the city. If the interest groups and actors are analysed on the basis of their members' position in the labour market, it should be taken into consideration that people employed in education, health, social and other services funded through the welfare state are a significant segment of the labour force. This may be especially significant in Scandinavian welfare states, where cities and other local authorities provide most of the welfare services. The total civilian public employment was about 37–38 per cent of the total employment in Sweden and Denmark and rising towards the same level in Norway and Finland in the 1980s. The same figure was about 24–25 per cent in France, Belgium and the UK and about 15 per cent in the USA (Therborn 1995). The majority of the Scandinavian public labour force is employed by the local authorities.

If the interest groups and actors are analysed on the basis of their members' position as consumers, the welfare state is also quite significant. First of all, the welfare state provides income for people who have lost their role in production, particularly, through old age and disability, and people living on unemployment benefits, social security and various other sorts of social/public income. In a normal Scandinavian city, these groups may account for almost a third of the inhabitants. About 13–19 per cent of the population are on old age pension, 4–7 per cent on disability pensions and 5–20 per cent of the labour force live on unemployment benefit (Nord 1994). They are not linked to society through the production of goods and services but they are still influential as consumers and beneficiaries of the welfare state. Social policies to support families with children and students in their education also mean that a significant part of the consumption of children and young people is related to the provisions of the welfare state. Children and young people under twenty years of age are 24–33 per cent of the population of an average Scandinavian city (Nord 1994).

A second link between the welfare state and the shaping of consumption-related actors is the fact that a significant proportion of consumption is through education, health, and social and other welfare state services. When consumption issues are the focus of political action, they are more likely to be issues related to welfare state services than to commercial services.

Differences in welfare states may be expected to lead to different impacts on the shaping of actors within cities. The large female participation in wage labour in the Scandinavian model is related to the great importance of trade unions, professional organisations and work-based groupings to female political action. The trade unions of the welfare state employees, such as the unions of nurses, teachers and employees of kindergartens, are major channels for female influence in Scandinavian local and national policy. This is also one of the reasons why women play a much greater role in public policies in Scandinavia than in other European countries. It is not exceptional for more than one third of the members of a local council to be women. This is also reflected in national public policy. More than one third of the members of Scandinavian parliaments and governments have been women.

Extensive and universal income transfer systems ensure a more consumption-oriented shaping of political interests for larger groups than just the middle classes. Thus it ensures a more significant role for the consumption-oriented shaping of political actors. In particular, pensioners are becoming major political actors in local politics. Their role within political parties seems to have increased: they are a significant fraction of the membership of most traditional political parties in Scandinavia and there have even been attempts to create parties particularly for pensioners. Pensioner organisations also form coalitions across party borders to influence local decision-making on issues such as reductions in public transport charges for pensioners and services for the elderly.

Industrial products may be stored for long periods and transported over long distances. Thus, production and consumption, as well as producers and consumers, may be separated with regard to both time and place. On the other hand, personal services have to be produced and consumed at the same time and in the same place. With regard to many services, such as child care, education or home nursing, consuming households are not only consumers but, at the same time, often co-producers. Good child care, primary education or home nursing can only be brought about by collaboration between the household and the welfare state service providers. This also means that when interest groups and other actors develop around service issues, they are quite often a mixture of producers and consumers. For instance, physicians and potential patients may join together to defend a hospital under a threat of closure or leisure service providers may join with their clients to support a common environmental interest.

The joining of welfare state service employees and their clients around welfare policy issues is of particular importance in Scandinavian cities. A significant number of new social movements in Scandinavian cities concern action around welfare state issues, such as defending a hospital from closure or demanding more child day care places or sports facilities. The role of the welfare state professionals is almost always central to these movements. The professionals influence city policies as: (1) expert advisers to the elected politicians; (2) members of their strong local trade unions which defend their interests in negotiations with the city as an employer; and (3) an important electorate with a voting power of more than one fourth of the whole electorate. By helping their clients or consumers to organise to defend the consumer interests they are able to create a fourth channel for influencing the decision-making in the city. From the perspective of the client or consumer movement, they are very important allies, because of their many influences over local decision-making. This, however, also means that consumer movements quite easily become dependent on the support of professionals.

Thus, it is no surprise that local welfare state employees are quite influential groups in Scandinavian local politics and as leading members of local trade unions, political parties, non-governmental organisations and social movements. This is also an important aspect of the resistance against pressures for welfare cuts in Scandinavia.

More comparative research is needed between different welfare states before drawing firm conclusions. However, from the Scandinavian perspective, it seems that the following conclusions might be made:

- The Scandinavian welfare state increases the opportunities for women to have impact on local authorities through their working life and labour-market-based organisations. In addition to autonomous radical feminist action, a kind of 'state feminism' is also reflected in the pattern of political action through trade unions, political parties and movements joining professionals and clients.
- The expansion of the welfare state has a significant impact on the shift from production-based shaping of political action towards a consumption-oriented shaping of political actors in local politics. In particular, the significance of pensioners in local politics is increasing. Consumption issues that are the focus of local political action are also often issues of welfare state services.
- There are great opportunities for the shaping of actors on the basis of a partnership between service employees and service consumers, thus combining the production-based and consumption-oriented shaping of political actors. Welfare state services form a particularly fertile ground for such a partnership. In the Scandinavian welfare state model, such a partnership has opportunities for the greatest political influence, but also includes a high risk of professional dominance within the partnership.

Much of the influence of the new, more consumption-oriented political interests have, until now, been channelled via the old political structures. The most important political parties in the large Scandinavian cities are still the traditional ones: social democrats and the urban bourgeois conservative parties. Local trade unions and employer and entrepreneur organisations are still significant power bases. While the social democrats and conservatives seldom co-operate in national governments, with the exception of Finland in 1987–91 and since 1995, there is more co-operation between these traditional urban parties in policy-making at city level. The traditional political groups are, however, challenged both from outside and from within. The green party is the third biggest group in the city council of Helsinki and is also significant in Stockholm. Feminist groups, environmental groups, pensioner groups and other new groups have succeeded in winning seats on city councils in all Scandinavian countries. Within traditional parties, tensions between public- and private-sector interests, between female and male interests, between generations and between other groupings have become more visible.

Cities as actors in the welfare state

Most of what has been written about welfare states, in particular from an international comparative perspective, has dealt with national welfare states. Cities and other local authorities have been viewed as being less important in the implementation of welfare policies.

There are, however, at least three issues that have been linked to a need or an opportunity for giving more emphasis to the local and, in particular, to the city level in social policy:

(1) the need for developing local responses to new local, particularly urban, social problems;
(2) the assumed weakening of the national state, as a result of the globalisation of the economy and West European political integration within the EU; and
(3) the conscious decentralisation within the public sector.

More local social problems?

Major social problems have often been linked to certain places. The history of many Scandinavian countries tells of problems such as the poverty of landless rural families, the problems of workers' communities developed at the boundaries of industrialising cities, the shortage of housing in expanding cities, the social imbalance in the areas of greatest emigration, or the social problems of the new suburbs.

Although the 'definitions' of the social problems refer to certain places, the policies to alleviate these problems have sooner or later developed into national social policies. As the problems have very often been understood either as causes for people to start moving to cities or as caused by rapid migration into cities, the underlying motive for social policy strategies seems often to have been the regulation of migration.

The underlying principles of the 'Scandinavian welfare state model' also lead to national solutions to localised problems. The principle of universalism does not allow for social policy that could be understood to lead to different social rights for inhabitants of different parts of the country or for members of different socio-economic groups. As institutional social policy, Scandinavian social policy may be contrasted to social policies implemented as particular programmes for particular population groups or particular social problems. Thus, even if the welfare state has had an enormous impact on the shaping of the social and spatial structures of Scandinavian cities, welfare state policies have never been defined as 'urban policies'.

The difference between the present and the earlier discourses on urban social problems is that now the problems are understood as genuinely urban, without a significant aspect of migration from rural to urban environments. For instance, while unemployment was previously often higher in areas of emigration to the cities, now the unemployment rates seem to remain higher in the biggest cities, such as Copenhagen, Stockholm and Helsinki.

The earlier phases in the development of the national welfare state were also strongly linked with the fact that trade unions and employer organisations concentrated their negotiations at the national level and drew the state into the negotiations – for instance, as part of a national incomes policy. Now the trend is towards more local negotiations and contracts between employers and employees. It is argued that this is needed in order to increase flexibility in employment, including atypical employment conditions. If the link between social policy and labour market bargaining is to be sustained, it may be argued that a more local approach to social policy is needed. This could help in compensating flexible labour market income and service needs (for instance the child day care needs of women in flexible employment) by locally flexible social policies. However, what is meant by 'local' in labour market policy is quite different from local in terms of cities and municipalities. There, local normally means the level of an enterprise or a production unit. Thus there may be a need to shift the focus of social policy and co-ordination between economic and social policy from the national level, but this does not seem to increase automatically the role of the cities in the welfare state. Quite probably it would create new problems for the co-ordination of social and economic policies.

A third argument for more local social policy could be the assumed greater economic differentiation between different regions and cities, as a result of a global restructuring of the economy (Lash and Urry 1994). It may be argued that social policy should be different, for instance, in industrial centres in crisis, in flourishing high-tech regions and in urban regions acting as centres for communication, transport, producer services and finance networks. Although policies on subsidies, and tax deductions and investment on infrastructures, have responded to these regional differences by differentiating benefits given to different regions, the differentiation of social policies by different cities or regions has not been the normal Scandinavian pattern. During the last decade or so, regional differentiation has forced Scandinavian countries to complement universal and institutional social policies by special central state funding for activation policies focused on areas of the highest unemployment or other problems. The Structural and Social Funds of the EU have also supported this focusing. This may be understood as a logical consequence of regional differentiation in economic policy. If there is to be co-ordination between economic and social policy and if co-ordination of economic policy is regionally differentiated, there is an increased need for regionally differentiated social policy as well. Social policies may also become more important assets in the competition between different cities and regions.

Competition between cities on private and public economic investments has a long tradition in Scandinavia. There has been competition for the central funds for infrastructure, such as roads, railways, harbours and airports. Competition for universities and other institutions of higher education has been fierce. Thus, good political and personal relationships between the central state and the cities have been important assets in competition. Cities also compete in the welfare services they offer for the employees of private enterprises. When unemployment was low and there was even a shortage of labour in many sectors of the economy, good provision of housing, child day care and education for young families were significant assets. Usually, the competitiveness of the capitals have been strong enough even without competing on welfare. Competition on welfare has more often been a strategy for smaller cities and for different satellite towns within the metropolitan areas. Thus, even in welfare states which emphasised universalism and regional equality, there was some room for competition on welfare policies.

Economic restructuring and high unemployment change the impact of welfare in competition between cities. Because there is no lack of labour in general but rather a need for a workforce with particular qualifications, policies on education become even more significant. Most Scandinavian cities still believe in high-quality cultural, social, health and education services as well as in low poverty rates as assets in competition. However, there are political pressures to allow greater income differentials and also to favour differentiation in the provision of welfare services, which would threaten the principle of universalism in welfare policies. Until now, the Scandinavian welfare state 'model' and its city-level structures and institutions have mainly remained stable and resisted the pressures for any basic changes.

Weaker national state?

A welfare state is part of a national state and the development of welfare states may be seen as the most recent phase in the long process of strengthening national states (Therborn 1995). The globalisation of the economy and the strategies of dominating enterprises mean that the boundaries between national states lose their significance. At the same time, the significance of subnational or cross-national urban regions in the creation of networks and socio-economic infrastructure for the competitiveness of enterprises emphasises the role of the regional level.

The development of supranational political structures and, in particular, of the European Union, can also be seen as a response to economic development with regard to social policy. Many pressures and means to 'harmonise' national social policies have been identified (Leibfried and Pierson 1995). At the same time, the main social policy programmes of the Union are implemented through the Structural and Social Funds, which support the development of regional approaches within member states.

It seems to be quite difficult to estimate the future impact of these supranationalisation and regionalisation tendencies with regard to the welfare state within the European Union. It seems quite probable, however, that the national state level will not be as dominant as it has been in the twentieth century. It is more difficult to say how much of what is now controlled by the national state will be moved to the supranational level and how much to the city or the region. It might be, for instance, that the basic social security benefits will slowly be harmonised at the supranational level, while the differences between the urban regions may increase in the provision of collective services and earnings-related additional social security benefits (Lehto 1997).

Decentralization of the welfare state?

Governments in many West European countries have given many promises and launched many initiatives to decentralise the public sector. The arguments for decentralisation have included promises both to increase democracy by moving decision-making nearer to the citizens and to increase effectiveness by introducing more flexible and less hierarchical management patterns.

It is, however, possible to ask to what extent there is a real process of decentralising in welfare states. The reforms quite often only deal with public services, and 'benefits in kind'. At the same time, the administration of the main social security benefits may be kept in the hands of the central state. The reforms are also introduced at a time of change from expansion to budget cuts. Thus, while decentralising some of the formal decision making, the 'dictatorship of scarcity' may be an even more powerful harmonising force than earlier bureaucratic central state guidance of the cities and other local authorities. For instance, in Denmark, during welfare cuts in the mid 1980s the decentralisation of public service management was paralleled by a diminishing role for the local authorities in public spending, and a similar development may be identified in Finland, and to a lesser extent in Sweden, in the early 1990s (Kautto *et al.* 1999). Finland, which has experienced the fastest expansion of the welfare state, also seems to be experiencing the fastest changes during the welfare cuts. If decentralisation/centralisation in Finland is measured by changes in the proportional roles of the municipalities and the central state in making decisions over the public finances, the early 1990s seems to be a period of quite dramatic centralisation.

Different decentralization in different welfare states

Although the supranationalisation, centralisation and decentralisation trends are, at least to some extent, similar in the whole of Western Europe, there are also differences. The Scandinavian welfare states, in small and quite homogenous countries, may be, at the same time, both quite centralised and decentralised. The cities have a large role and significant autonomy but, at the same time, large centrally defined responsibilities in the provision of welfare state services. Social policy that emphasises universalism and does not accept large regional differences has demanded strong national institutions that guarantee the same social security and similar services all over the country. The political party and interest organisation structures in Scandinavia have also been much the same at the city and national levels. Thus, there has not been too much tension between central government and local government, although these tensions have increased somewhat due to cuts in public spending and in subsidies to local authorities.

Larger and more federal European countries, including Germany, Italy and Spain, tend to have greater differences between the regions. A welfare state with weak local government may also lead to greater regional variation in the absence of strong local pressure for an equal distribution of welfare state resources.

Whatever the relationship between the national and the city level in social policy, it seems that the growing significance of the cities in global economic restructuring leads to an increasing need for partnership between the local and the national and – in a growing number of issues –

also with the supranational levels in making social policy. The more important role of the cities in social policy does not mean that the national role would decrease or that the cities would be 'freer' from the pressures of the global economy. The growing significance of the cities is better described as an increasing need for negotiating and creating partnerships, both between the different actors at the local level and with the different actors at a national and supranational level. Most large Scandinavian cities seem to be in a process of developing the necessary new capabilities, partnerships and strategies for a proactive role in the restructuring of the economy and the welfare state.

Concluding remarks

The European welfare states are under strong economic and political pressure at present. Many observers call the situation 'a crisis of the welfare state'. The welfare states did not succeed in preventing unemployment rates from growing and this deepens financial problems for welfare policies. The institutional structures developed before the present labour market restructuring do not produce equality as effectively as before. Thus there may be significant changes in European welfare state policies. However, the institutional, economic, political and ideological differences between welfare states will probably remain significant.

This chapter has discussed the impact of the welfare state in the shaping of cities, in the shaping of actors within cities and in the shaping of the role of the city in social policy. It has been mentioned that cities that developed in the absence of the welfare state differ from cities that developed during the period of the expansion of the welfare state. It is also quite probable that cities that will develop during the period of the restructuring of the welfare state will also differ from present-day cities. It has also been noted that different welfare states shape their cities in different ways. Thus, the differences between US and European cities, as well as between Scandinavian and central European cities are, at least partly, a result of the differences in their welfare states.

The welfare state, by influencing the development of social and family structure, by distancing the opportunities for consumption from one's position in production and by creating new issues and new opportunities for partnerships between the producers and consumers of welfare state services, has a significant impact on the shaping of interest groups and political actors within cities. It is argued that this impact is greatest in the Scandinavian welfare state model, but it should not be underestimated in other welfare states either.

Global economic restructuring challenges the capacity of the national state to regulate developments in the economy as well as to exert political power over other issues. Both supranationalisation and localisation of political power are predicted. This could also increase the role of local authorities, and in particular of cities, in social policy. However, the localisation and decentralisation tendencies in social policy seem to be rather contradictory. What seems to be obvious is a growing need for cities to negotiate and create partnerships in social policy between different local, national and supranational actors. This may not mean increasing decentralisation or local autonomy, but may be better described as a need for better co-ordination and better adaptation to increasing regional and local differences in the globalising economy.

6 Social movements in European cities: transitions from the 1970s to the 1990s

Margit Mayer

Urban social movements reveal that cities are not unified political actors: their protest activities and the demands which they insert into the political arena reveal cleavages and conflicts within the local space that have consequences for how cities may act on the larger stage.

This chapter looks at how urban social movements have developed over the past few decades in order to assess their current shape and significance for European cities. It interprets the movements in the context of structural trends, which have taken somewhat different forms in the different European countries as well as different forms in various types of cities. Depending on the differing degrees of modernisation, of centralisation and of decentralisation, the strength or weakness of political parties able to absorb and process the disruption and conflict, and the openness or repressiveness of the given political opportunity structure, outcomes in the different countries may range from intense movement mobilisation and interest group activity to no extra-parliamentarian activity at all. More detailed country-specific analysis is necessary to explain such varying patterns, and to analyse the role movements are playing in the formation of cities as political actors. This chapter is but a first step in capturing some significant shifts in the make-up and role of urban social movements from the 1970s and early 1980s, when these movements exhibited relative coherence and unity in their opposition to urban renewal, in their demands for improved collective consumption, and in their challenge of the established parties' and local governments' monopoly to process political interests (cf. Castells 1973; Ceccarelli 1982), to the 1990s, which have seen an extremely fragmented urban social movement scene all over Europe. The chapter draws most of its material from the (West) German situation, though comparable developments in Britain, France and other European countries will be noted.

The late 1960s and 1970s: broad coalitions and politicised opposition

The post-war era saw its first massive phase of urban movements at the end of the 1960s and during the early 1970s, when citizens' initiatives mobilised against large-scale renewal projects and in defence of residents' living conditions. These struggles soon expanded into struggles over the cost and the use-value of the public infrastructure in which cities had begun to invest, and created a fertile milieu for various types of grassroots and community groups into which New Left students and their projects inserted themselves and managed to provide some ideological coherence.

An important background to this wave of mobilisation was the expansion of the (social-democratic) Keynesian welfare state model of development, which became the imperative all over Europe from the mid 1960s onwards. Local governments implemented this scheme by expanding the urban social and technical infrastructure, by organising the provision of land for urban development and by managing large-scale urban renewal. All over Europe, and not merely in places governed by social democrats, large-scale urban renewal and modern housing construction were at the core of local politics. This type of urban infrastructure and of collective consumption expansion accelerated the segmentation of urban space into monofunctional zones of residence, shopping, working and entertainment, thereby frequently destroying the vital fabrics and milieux of neighbourhoods. While serving to raise consumption levels, it also standardised ways of living and monotonised urban life. It was against the effects of these growth strategies that the first phase of urban oppositional movements during the 1960s and the beginning of the 1970s mobilised.

Citizens' groups initially used conventional, pragmatic methods to defend their neighbourhoods and chose co-operative tactics and professional strategies such as 'planning alternatives from below'. However, where they confronted unresponsive technocratic city administrations they would resort to more unconventional forms of politics, including direct action and street protest. Contested issues included not only infrastructure expansion, but also the cost, quality and participation in its design. In many cities, broad mobilisations occurred which were directed toward lowering costs and influencing cultural norms expressed in the institutions of collective consumption (especially schools, kindergartens, and public transport). Often, these protests were joined by initiatives from the youth protest movements, whose roots were in the 1960s' anti-authoritarian movement, and by New Left groups. In the course of the 1970s, waves of leftist community-organising groups 'infiltrated' these local movements in Italian, French and German cities. The New Left saw the 'reproductive sector' as an area of politicisation where disadvantaged groups could be mobilised, and their analyses were translated and disseminated across the leftist scenes of Italy, France and Germany (cf. Cherki and Mehl 1978; Cherki and Wieviorka 1978; Godts 1978).

During the first half of the 1970s, squattings took place in Dutch, British, Danish and West German cities (cf. Bodenschatz *et al.* 1983; Scudo 1978). Among the most favourable conditions for a concentrated and relatively long period of squatting were those in Frankfurt, where from 1972 to 1974 squatting actually dominated the local movement sector. The issue here was the restructuring of a neighbourhood near the central business district for expanding tertiary functions. In other cities the triggering events were large-scale renewal projects. In each case, renewal plans and large-scale demolition of turn-of-the-century housing did not include open and democratic planning processes, but instead non-public and generous deals between city agencies and investors. This political process stimulated speculative behaviour whereby whole blocks would be bought up, temporarily rented out to 'transitory residents' like students and immigrant workers, or left vacant. The realestate owners could expect huge profits from the demolition and eventual construction of high-rise office buildings. Both the undemocratic nature and the detrimental social effects of such urban development politics sparked (in Frankfurt as early as 1968) protest organised by conventional citizens' associations, which usually saw themselves as non-ideological and quite distinct from the radical, student-led groups which were also forming at the time.

Within this climate the first squat in Frankfurt was carried out by students and social workers who had already been active in community groups. The squatters had formed a 'collective living experiment' and occupied a large, turn-of-the-century building together with Italian immigrant families. Two similar squats followed a month later and their success bred imitators in other circles (Dackweiler *et al.* 1990: 210). In Hamburg, the first squat took place in 1973 in a similar atmosphere of widespread protest against demolition plans (Schubert 1990: 35). Explicitly political projects, the goal of these first squats was to radicalise political work in the 'reproductive sphere'. The squatted houses both symbolised the criticism of urban renewal that consisted in demolition for luxury housing or offices, and served as organisational bases for further squats; their residents also played important roles in initiating other movement activities.

Because the issue of urban destruction was easily presented as a political scandal, public reactions were initially quite positive.[1] The occupation and subsequent violent eviction of a building in September 1971 encouraged more squats, because widespread indignation over the brutal police actions and bloody street battles forced the Frankfurt mayor to rescind his earlier eviction order. Similar sympathies arose in Hamburg over the city government's repressive and criminalising response to their first squattings. Citizens' initiatives, tenant groups and professionals came to the support of the squatters and formed a broad housing movement. In Frankfurt, from October 1971 to July 1972 ten more mostly successful squats took place, broadening the infrastructure for political work and the movement's alternative living arrangements. During this expansive phase of the squatting movement, a curious coexistence and even productive relationship prevailed between the radical, anti-reformist protest and social-democratic reform policies, which attempted innovative and socially responsible solutions to the problem. The lines of conflict were drawn between the squatters, their supporters and the ruling SPD against what appeared to all as the common enemy: the speculators and irresponsible real-estate owners.

Reasons for the decline of this first wave in the squatting movement differed according to the local situation. While in Hamburg or Berlin community and tenant initiatives worked pragmatically to prevent demolitions and to create and maintain alternative housing forms (cf. Bodenschatz *et al.* 1983), thus building an organisational basis for another massive mobilisation during the early 1980s, in Frankfurt the movement's strong infusion of political and existential radicalism eventually turned into a limitation. Left-wing radicalism and militancy became quite synonymous, both because of the strong presence of New Left groups within the movement, who understood their activities in the reproductive sector as part of broader revolutionary activities, such as party-building or internationalism, and also because of the SPD city government's changing political strategy. In 1973/4, the city began urging evictions while simultaneously presenting itself as the saviour of the existing housing stock and the fabric of the threatened neighbourhood (Dackweiler *et al.* 1990: 214). The stiff repression and criminalisation of the squatters, during two protracted eviction conflicts in particular, intensified the movement's critique of SPD reformism and its own self-radicalisation (Stracke 1980) while the distance from the more moderate citizens' initiatives increased and the supportive environment began to crumble.

In spite of such setbacks this phase produced, by the mid 1970s, in most European cities a new political actor: a self-confident urban counter-culture with its own infrastructure of newspapers, self-managed collectives and housing co-operatives, feminist groups, etc., which prepared to intervene in local and broader politics.

The world-wide economic recession of 1973/74 indicated a break with the post-war growth model. Markets for consumer durables and mass products had become saturated, the labour process could not be further taylorised, economists noted a structural crisis of capital reproduction, and the consensus around Keynesian policies dissolved. In other words, the social and technical limits of the Fordist growth model had become apparent: the rigidities of the production structure, the rising costs of mass production and mass consumption, and the politicisation of those costs and effects slowed down growth rates and triggered social conflicts and new social movements which put these costs on the agenda. Citizens' initiatives protesting the threats and infringements to their living conditions contributed to making the social limits of the Fordist regime visible, particularly how its resource- and waste-intensity creates barriers to expansion. The ecology and other movements also challenged technological fixes as solutions to the Fordist relationship to nature, fearing their negative effects on democracy and social responsibility.

The economic restructuring efforts undertaken to overcome this crisis of Fordism augmented the proportion of social groups that remained excluded from the 'blessings of Fordism' as unemployment rates (affecting especially the younger age groups) began to skyrocket. Negative effects on urban living conditions encouraged not only more protest activity but also, particularly in West Germany, a coming together of heterogeneous movements and their creation of independent organisational structures quite opposed to the state and its parties. As repressive and marginalising measures by the authorities provided repeated cause for co-operation among the movement groups and for confrontation with the authorities, alternative projects, frustrated citizens' initiatives and new local social movement campaigns (peace, women, ecology) developed tighter solidarities and a shared radical-oppositional self-image.

In most European countries during the late 1970s alternative projects and communal experiments came to the fore more strongly: projects in all types of production and service activities and collective living arrangements were initiated (cf. Bertels and Nottenbohm 1983). At the same time, the continued experience of political exclusion and marginalisation during this phase led the movement groups, in Germany more than in other countries where progressive left-wing parties open to the ecological issue were in place, to shift their political interventionism in the direction of electoral alternatives.

Thus, even while the movements were still strong and had palpable impacts on their respective cities, a number of indications pointing to their later fragmentation had already become visible: the pursuit of different interests, the beginnings of the electoral route via ecological parties, and openings by the local state structures (in Holland, France or England more so than in Germany at this stage). The next section illustrates one decisive source of the eventual breaking up of the urban movement scene, which, though heterogeneous in its constituency, was held together by intense links, both substantive and formal, during this early phase.

The 1980s: shifts in the relationship between the movements and the state

The beginning of the 1980s saw another massive wave of squatting and housing struggles, but the surrounding context had changed. Local governments, forced to find new and alternative ways of dealing with the fiscal restrictions imposed by the consequences of economic restructuring, growing unemployment and rising welfare costs, began to look to community groups and alternative organisations for their innovative potential. Thus, in the course of the decade, a transition from urban social movements challenging the state to a less oppositional relationship between 'interest groups' and a local (welfare) bureaucracy increasingly confronted with its own limitations occurred.

The squatting movement of the early 1980s was soon known as the 'rehab squatting movement'. This time, the movement was strongest in Berlin, where it started in 1979 as the last desperate step of a ten-year-long defensive community and tenant-organising endeavour to stop the deterioration, forced vacancies and speculation carried out by private landlords. When a powerful youth and alternative movement emerged and coalesced with local community groups, squatting became a form of self-help in which the squatters not only occupied vacant buildings, but also attempted to restore the properties to liveable condition after years of physical deterioration. Again, these forms of occupation managed to attract the support of broad sectors of the population alienated by the corrupt building policies of the Berlin government and by the disruptive effects of – and huge profits made by – massive housing development, real estate and tax shelter syndicating firms. During the movement's peak in 1981, about 160 large tenement buildings were 'rehab-squatted' in West Berlin, directly involving about 5,000 people.

There were simultaneous widespread squatting movements in Zurich (cf. Kriesi 1984) and Amsterdam (cf. Narr *et al.* 1981), as well as other German cities related to the 'new housing needs', a term coined for what appeared to be a new problem at the beginning of the 1980s. Typical of this new housing need was also the more limited squatting movement in Frankfurt at the time, which was mostly carried out by younger radicals at the margin of the 'established' movement scene, many of whom were primarily interested in cheap housing. The local opportunity structures were no longer conducive to their efforts as the Christian democrat city government would not tolerate illegal occupations. Intense gentrification processes in the central business districts had displaced those groups who might have been willing to support and use radical forms of self-help. The dominant movement issues of the period, the struggle against the airport expansion and the new peace movement, did not leave much space for other mobilisations, and the activists of the earlier housing struggles had meanwhile gone into green electoral politics (cf. Dackweiler *et al.* 1990: 206 ff, 219 ff).

In Berlin, the joint actions of the squatter movements still brought together distinct groups with distinct interests: citizens' and tenants' initiatives, marginalised youth, and alternative political groups. While the former were interested in careful urban renewal and self-help in housing rehabilitation, the latter sought niches for themselves in a relatively protected milieu, used the actions as a stage for their struggle against the state, or were simply interested in suitable space for political projects. While initially important bonds were connecting them – a radical critique of the state housing policy and a desire for unfettered self-realisation, for private spheres without state control – these eroded as the fruits of their self-help labour were repeatedly destroyed by evictions, demolitions, and drawn-out court cases. Eventually, some squatters and support groups began to work up proposals for the transfer of squatted houses into public ownership, 'legalised' self-management and long-term leaseholds, as well as an institutionalised third-party mediator and manager between the houses and the state. Berlin spearheaded this development, which, after years of struggle and many setbacks, gradually splintered the movement, producing an alternative renewal agent (*Stattbau*) that was to administer the buildings on behalf of the Berlin Senate, which would in turn purchase the buildings from their owners and give squatters long-term leases with extensive self-management rights (cf. Clarke and Mayer 1986: 412). Following this model, similar alternative renewal agents were established in Hamburg in 1984

(Schubert 1990) and over the next few years in other West German cities.

A similar process of 'approximation' took place with the alternative collectives and citizens' initiatives and the state. In Berlin these groups had formed an umbrella organisation 'Arbeitskreis Staatsknete' to secure public funding for their projects. And while the founding activists among them framed this demand as a political offensive on the 'new voluntarism' propagated by the christian democratic government, more and more projects joined the Arbeitskreis, that were new and had little political experience, but high hopes for *individual* funding. This changing composition among the activists reflected the fact that deteriorating economic conditions and increasing marginalisation (especially youth unemployment) were beginning to undermine the position of alternative projects all over Germany (cf. Beywl 1983: 97; 1988). A consequence was that the projects sought to professionalise and were increasingly willing to participate in the political bargaining process wherever it would open up to them.

Thus, both housing movements and alternative projects on the way to becoming alternative service delivery agencies began to receive funding and some were even incorporated as 'model projects' into municipal social or employment programmes. Their formerly antagonistic relationship to the (local) state had given way to a stance of working 'within and against' the state; the dynamic of this new work also, as we shall see in the next section, set these types of movement groups apart from others that continued to mobilise outside of a routinised co-operation with the state.

The 1990s: a fragmented movement scene

While the various kinds of struggles described for the last two phases still go on, the forms of urban protest have become differentiated and more separate from each other. Since the late 1980s, the social composition and the political orientation of the urban movement milieux have become increasingly heterogeneous, manifesting more and more polarisations, cleavages and also forms of implosion. There is limited overlap or co-operation between middle-class citizens' initiatives defending the quality of life of their neighbourhood and new poor people's movements; the community-based and alternative groups that have become inserted into innovative municipal programmes now even find themselves attacked by radical 'autonomous' protest groups; and campaigns against large urban development projects often appear to operate independently of any of these groups making up the movement sectors of most European cities. This section looks at the urban social movements that seem most active today[2] and inquires not only into their political demands but also their contribution to the newly emerging forms of urban governance.

One of my arguments is that all of the protest activities currently visible respond to trends of restructuring, which characterise cities' trajectories in the 1990s. Relating the urban movements to these larger trends in urban restructuring will allow us better to assess their potentials and their likely perspectives.

The not-so-NIMBY NIMBY groups: struggles in defence of the community[3]

The most widely studied segment of the current urban social movement scene is that of frequently middle-class-based, quality-of-life-oriented movements focused on protecting their home environments – from too much traffic or development. Where intra-urban competition has intensified, municipalities' increasing concern for economic development has frequently been translated into policies favouring increased automobile traffic or subsidies for plants with hazardous emissions or other negative effects for surrounding neighbourhoods, triggering protest and resistance. Case studies show that such groups quickly become skilled at a variety of tactics and repertoires such as petition drives, political lobbying, street confrontations and legal proceedings.

Researchers tend to lament the fact that social justice orientations, which used to characterize the goals and practice of such citizens' initiatives during the 1970s, have been replaced by particularist interests and/ or a defence of privileged conditions[4] (e.g. Krämer-Badoni 1990; Krämer-Badoni and Söffler 1994).

But not all of these environmental or quality-of-life initiatives mobilise for their private and basically exclusive interests. There are also case studies of local movements composed of working-class and middle-class participants mobilising against highway construction plans, traffic congestion, or housing shortages, and against polluting industries and hazardous facilities, with which poor/working-class communities tend to be disproportionately burdened. They demand not only tighter controls and a reduction of toxic emissions, but also representation on relevant decision-making boards and they tend to formulate more universalistic and less parochial claims.

'Protecting the home environment' can thus take ideologically rather divergent forms. Whether such movements develop more towards privilege protection and exclusiveness or maintain progressive agendas seems to depend less on their (middle- or working-) class composition, but rather on their particular resource environment, as mobilisations around taxes have dramatically illustrated. The anti-poll tax movement in Britain of 1989-90 (cf. Hoggett and Burns 1991-92; Burns 1992; Bagguley 1993) contrasts in an interesting way with the home-owners' movements in US states such as California or Massachusetts: both have raised questions of justice and community control, both constitute territorial mobilisations in defence of home values and neighbourhood stability. While the composition of the home-owners' movement was lower and middle class, in fact it provided the grassroots resources for a property tax revolt led eventually by wealthy property-owners and commercial interests (Fainstein and Hirst 1994) culminating in the campaign for Proposition 13 in 1978 (Lo 1990). The British campaign, however, got its resources from established political activists, often from the public-sector middle class, left-wing organisations and local tenants' groups and churches, which allowed it to maintain its natural justice/moral economy orientations.

Struggles against the new politics of urban development

The new strategies of urban revitalisation and redevelopment – shaped by intraregional competition and turning increasingly towards 'large projects' – have also triggered a specific kind of battle: one over the design of the city overall. Large projects or festivals – such as the Olympics, World Expo, international garden shows or 1,000-year anniversaries of the granting of city charters – have become instruments of urban politics, against which opposition movements have mobilised, arguing the detrimental side effects of and the lack of democratic participation inherent in these strategies of raising funds and of restructuring the city. Similarly, spectacular urban development projects (such as London's Docklands or Berlin's Potsdamer Platz) are criticised for their spatial and temporal concentration, preventing any salutary effects being felt by the city as a whole (cf. Häußermann and Siebel 1993). Furthermore, the campaigns to attract 'mega-events', sports-entertainment complexes and theme-enhanced urban entertainment centres depend on the packaging and sale of urban place images.

Against such politics, broad coalitions of local movement groups have come together, joining actors and political positions who otherwise do not have much overlap, as, for example, in the NOlympia Campaign in Berlin in 1991-3 or in the opposition to the Hanover Expo 2000 (cf. Selle 1994). The potential of these movements, which tend to go beyond particular community interests and which raise questions of democratic planning that urban elites concerned with interregional and international competitiveness would like to downplay,[5] has yet to be studied.

Routinized co-operation with the local state

Unlike the first two kinds of protest that mobilise against a local NIMBY movement or project of urban development and that are usually characterised by precarious, if wide-ranging funding arrangements, a third kind is more or less institutionalised and benefits primarily from municipal funding programmes, supported occasionally by state-wide, national or EU funding sources. These programmes are part of the new strategic urban revitalisation efforts as well as of the attempt to address the state apparatus' difficulties with producing efficient welfare.

This kind of movement involves projects and initiatives whose demands for rehab housing, community economic development, client-based social services or women's centres have been acknowledged by the local state in a form that seeks to tie the movement organisations and their labour into the fulfilment of these demands. As was shown in the last section, the establishment of alternative renewal agents and equity programmes, and the funding of self-help and social service groups nowhere occurred overnight or without friction, and there has been interesting cross-national diffusion at work (see the ongoing work comparing such approaches across Europe in Froessler *et al.* 1994), but by the mid 1980s even the comparatively autonomous and anti-state German movement projects had begun to receive funding and some were even incorporated as 'model projects' into municipal social and employment programmes (Mayer 1993). Since then, community and/or movement groups' participation in different policy sectors has become routinised (cf. Froessler *et al.* 1994; McArthur 1995).[6]

Even in the former GDR the citizens' movements of 1989 were soon displaced by social movements that rapidly reached the level of formalisation and professionalisation which Western groups had reached more slowly. Old and new movement groups adapted rapidly to the new political and institutional conditions: they acquired the formal structures necessary in order to qualify for funds, sought contact and coalition with other groups in order to obtain information and not be isolated from negotiations with local administrations, and developed the necessary internal specialisation to qualify for the massive public funding for such groups as part of a nationally conceived labour market instrument for eastern Germany after 1991 (Blattert, Rink and Rucht 1994).

The sources for this co-operation between social movement organisations, state agencies, and established associations lay with the limits of each: the projects gained stability and security from making themselves recognised partners within the housing and social policy networks; municipalities opened up to the voluntary sector in order to address the crisis of the welfare state; and the large, traditional welfare associations began to absorb the innovative strategies of community-based organisations in order to overcome their rigid service structures.

The bulk of the research focusing on these novel forms of institutionalisation of social movements emphasises the contestatory character of their constituency and the counter-weight they pose to conventional ways of local economic planning and service delivery. Whether in the economic development sector, the field of alternative services, or that of women's projects, the work of the groups is generally found to be an innovative and progressive challenge to public policy, as improving access to the local political system and providing potentially more active citizenship (cf. Selle 1991; Jacobs 1992; Froessler *et al.* 1994).

However, more detailed studies show that this institutionalisation of community-based groups via municipal support is fundamentally ambivalent: while these groups do indeed contribute to the stability of the movement infrastructure and thus to the conditions for continuing mobilisation, their own democratic substance is far from guaranteed. They are subject to the danger of institutional integration, 'NGOisation', and of pursuing 'insider interests' (Roth 1994; Fehse 1995; Lang 1995). On the one hand, the institutionalisation processes have led to new intermediary, publicly financed or directly municipal agencies (such as houses for battered women, multicultural offices, self-help bureaux) so that movement milieux now include associations and public institutions that are not characterised by protest politics but provide infrastructural stability; on the other hand, this widening and increasing internal differentiation of the movement scene has led to growing conflicts and antagonisms within the movement sector. For example, alternative renewal agents and community-based development organisations, who are busy developing low-income housing or training and employment opportunities for underprivileged groups, find themselves criticised and attacked by other movement actors who do not qualify for the waiting lists or who prefer to squat (see the next two subsections). A further argument against the overvaluation of these groups' progressive potential are the effects of cities' exacerbating fiscal crisis, which has meant that public-sector funding for these groups has in many cases been drastically reduced, and claims to participation from these local movement milieux are increasingly rejected as a 'no longer affordable luxury'. The impacts of this renewed precariousness have largely been ignored by social movement research (cf. Mayer 1996).

Radical and 'autonomous' protest

A leading actor in the NOlympia campaign against Berlin's plans to host the Olympics in the year 2000 and the resulting urban restructuring has been the 'Autonomous' who, in many German cities, form a radical current of broader local mobilisations. There are similar left-wing radical groups active in French, British, Dutch and Danish cities. They mobilise not just against large projects and mega-festivals, but also against gentrification, displacement and highway expansion, and the increased policing and surveillance of public space – i.e. against those forms of urban renewal that are about to destroy the urban milieu they still thrive in (though they are framing this mobilisation in terms of a struggle against 'capital and the state').

They seize on the importance image politics have gained for the development of cities and seek to devise image-damaging actions – for example, in the course of the NOlympia campaign – to make cities less attractive to big investors and speculators. Their action repertoire is broad, ranging from direct action and squatting to uncovering and publicising the plans and methods of large developers and speculators, causing a scandal over the production of new poverty and homelessness, all the way to sometimes militant attacks on 'yuppies' carrying out gentrification, or on renewal agents (local officers working on urban regeneration projects) but since the mid 1980s they have also been attacking those who seem to work too closely with the authorities. Not only do renewal agents and incorporated citizens' boards seem 'established' to them, and appear to contribute, if unintentionally, to preparing the way for gentrifiers to move into the neighbourhood, but since the mid 1980s they have also been attacking what seems to them an integrationist strategy pursued by the 'Realo Greens' (cf. Lauterbach 1994: 111). This illustrates the division and polarisation within the movement milieu, which may be interpreted, in part, as a generational issue, but also as a product of the restructuring urban polity that includes some but not others in its new governance arrangements.

Protest by the marginalised: new poor people's movements

This last variety of protest arises not so much from within the movement milieu but directly from newly marginalised groups such as the homeless. The growing numbers of homeless people in all the Western metropolises are the most visible manifestation of new forms of exclusion and of the shift in social policies. Everywhere cities engage in futile efforts to 'clean' their central business districts, parks, train stations and subways. While sweeps, evictions and policing are being stepped up, the visibility and intensity of 'the problem' continues to grow (Mayer, Jahn and Sambale 1995). But there is a battle over people's willingness to accept this new degree of housing poverty and such measures of dealing with its 'victims' as a permanent fixture of urban life in the 1990s.

Homeless people, sometimes supported by advocates such as church groups, social workers or celebrities (as in Paris: see Body-Gendrot 1995), struggle against efforts to drive them out of city centres, occupy city halls, set up encampments, hold public forums, make demands on the city – and, occasionally, in the process develop solidarity, political consciousness and an organisational infrastructure, i.e. the elements which most social movement researchers consider pre-conditions for a mobilisation to arise.

Here, social movement research findings are scarcest since most authors assume this population to be not just poor and without resources but also disempowered and passive. There is evidence, however, that suggests that this is simply not true, leading us to examine our scientific assumptions as well as asking what it is that is emerging here. I am not aware of a European study that looked at homeless protesters over time, but there is a very interesting American study of homeless demonstrators (David Wagner 1993) that found how the participants of a 'tent city', about 100 people in Portland, Maine, were empowered by their action and were also able to achieve significant concessions from the city. Other localised tent cities or other struggles by newly marginalised groups have been less successful, but in any case, the emergence of this type of actor who – under certain conditions – will participate in social movement actions to put pressure on city administrations, needs to be followed up not only because it underlines the new structural realities, but because it promises a new view into activism and activists.

German marginal cultures have produced a distinct kind of movement in this context which is influenced by the alternative culture of the last decade: the so-called 'Wagenburgen', i.e. groups of people squatting on vacant land, living in trailers, circus wagons or other mobile structures, who see their action as 'a form of resistance against the political, social and economic relations in this city and this country' (Vogelfrai 1994).[7] There are about seventy to eighty such sites in Germany (cf. Knorr-Siedow and Willmer 1994); of the fifteen in Berlin (housing about 300 people) twelve were within the city centre area and thus threatened by eviction or recently evicted.[8] Their political orientations cover a wide spectrum: while some use the freedom this lifestyle allows them for politically virulent activism (such as housing illegal refugees), others are content to explore alternative ways of living.[9] But evictions or the threat of evictions have brought them together in campaigns to pressure city governments to tolerate the sites, delay construction, or provide other acceptable locations. In Berlin, house squatters and 'Wagenburgen' – who otherwise do not co-operate with each other – have joined in an action week as both saw themselves affected by the Senate's decision to eliminate unconventional housing from central Berlin as it prepares to become the site of the national capital.[10] While after the first eviction (in 1993) support from church representatives and the movement scene was still felt,[11] by 1996 this milieu had become so fragmented that there was hardly a protest against the removal of the largest (and most problematic) of the central city Wagenburgen: when the East Side Gallery (housing about 130 people) was dissolved in July, residents at an alternative site at the north-west edge of the city mobilised resistance so that most of them have become homeless (*Tageszeitung*, 18 July 1996 and 19 July 1996; *Tagesspiegel*, 24 July 1996).

The distinction between these five types of urban battles is obviously not always clear-cut; occasional campaigns will bring both militant activists of the fourth group and advocacy planners (frequently found in the second) together with the marginalised groups mentioned last. Also, in more and more cities middle-class people are finding new ways to organise social movements 'against poverty and isolation' as, for example, with Hamburg's 'donation parliament', which collects contributions and democratically decides which homeless and alternative projects to subsidise (*Tageszeitung*, 14 February 1996: 10). And there are also longstanding organisations that are hybrids combining characteristics of different groups. The point is that today's urban social movement scene does not consist merely or even primarily of one kind of movement, and that it is socially and politically far more heterogeneous and internally fragmented (into insiders and outsiders even) than only a decade ago. The different fragments, however, each manifest more and less progressive variants and show marks of the new social movement heritage of the 1970s. If we want to account for the dynamic, the tensions and the continuing (if ambivalent) vitality of these different movements, we cannot only turn to social movement research. Its frameworks of cyclical rise and decline of mobilisation, 'natural death' of a movement through institutionalisation, or concepts of gradual differentiation of movement sectors do not grasp the transformations which social movements in European cities have undergone in the last two decades. We need to turn, therefore, to the larger trends affecting urban structure and urban policies in this era. These also help us to define what opportunities for effective social movement activism might exist today.

The role of urban social movements in European cities of the 1990s

Cities now play a more important role in the search for economic, social and political arrangements that are more adequate to internationalised competition and its consequences. While multinational firms can switch their commitments and investments, localities do have some negotiating power, because the concrete supply-side conditions are increasingly shaped at subnational levels of decision-making. Local governments function as *ad hoc* mechanisms to flexibly adapt to the changing conditions of competition; they do so with active economic interventionism and by organising local politics in partnership with an extended range of non-governmental stake-holders. Besides the more competitive and entrepreneurial forms of urban development and an expanding urban political system, a third trend is becoming increasingly characteristic of the current structural transformation and political reorientations: the erosion of welfare rights and the increasing social inequalities. While European cities are certainly nowhere near the US situation of deeply polarised cities, where spatial concentrations of poverty and unemployment have bred forms of social exclusion that appear as an 'urban underclass', unemployment rates in European cities have also climbed more steeply than outside of cities and the restructuring of welfare states has also led to new processes of exclusion (cf. Mingione 1993). Each of these three trends – the entrepreneurialisation of the local state, the expanding system of local governance, and the erosion of welfare rights – has influenced the dynamic of the urban social movement field. I will summarise this influence in this last section and draw some conclusions about the constraints and opportunities this has created for the movements' contribution to the current urban scenario and to the process of governance.

1. Contemporary forms of urban growth and development have generated specific areas of conflict. Intensified global competition, the weakening of nationally directed efforts at territorial equalisation, and the new tools and strategies employed by cities to compete on this level, have led to a new hierarchy between cities and eroded the

homogeneous concept of the Fordist city. Cities at the vulnerable end of this hierarchy have applied strategies to upgrade their locality in the international competition for investors, advanced services, or mega-projects, often with unwanted effects for resident populations. Tertiary development in the central business districts and new infrastructure projects, if not carefully implemented with an eye to their social effects, may lead to gentrification and displacement, congestion and pollution, and new forms of spatial segmentation (cf. Cattacin 1994). Opposition movements, which have either built on existing (latent) networks and organisations or have sprung up anew, range from defensive and pragmatic efforts to save existing privileges (which in some instances have been selfish, anti-immigrant or racist), to highly politicised and militant struggles over whose city it is supposed to be (as in anti-gentrification struggles or movements against other growth policies). Given that much of the new productivity depends on the quality of mental labour, and that therefore the quality of the reproductive process has become an important criterion in the competitive process, movements focused on collective consumption and 'quality of life' have frequently contributed to generate those conditions of productivity required by the 'informational labour process performed in the space of flows' (Castells 1989: 206). In such cases, the interest in place-based living conditions may unite movements with city governments. On the other hand, large metropolitan areas occasionally provide the backdrop for strong, middle-class-based quality-of-life movements to succeed in averting an unwanted facility with the effect that then a poor/minority neighbourhood is targeted. More often than not different neighbourhoods will be played out against each other; the conditions under which this can be avoided need to be researched. Certainly, the frequently favoured strategy of cities to compete via mega-projects and festivalisation does provide opportunities for city-wide coalitions and thus for overcoming some of the fragmentation that has become the norm.

2. At the same time, the movement terrain is also shaped by the opening up of the expanding urban political system and the strategy of many municipalities to employ former social movement organisations in the development and implementation of ('alternative') social services, housing, or economic development. The social movement organisations which have inserted themselves into these funding and implementation structures, while finding themselves increasingly threatened by cuts and the reorientation of state programmes towards making labour markets more flexible, play a complicated role within the urban social movement scene. On the one hand they

enhance organisation building and lend stability to the urban social movement sector; on the other they contribute to the fragmentation and polarisation of that movement scene: as a result, we witness segmentation and competition within the movement for funding (especially in times of fiscal crisis), and we get private lobbying strategies to secure jobs and finances instead of public pressure and therewith presence in the public discourses of civil society. The inclusion of community groups in revitalisation partnerships has meant that they become tied up with managing the employment and housing problems of groups whose exclusion by normal market mechanisms might otherwise threaten disintegration. For successful projects, this management of scarcity may turn into privileging them over others, similarly disenfranchised but still dependent on the exclusionary mechanisms of the normal labour and housing markets.

3. Increasing unemployment and poverty, coupled with the erosion of welfare rights and the shift in social policies, have led to the emergence of movements that primarily consist of disruption to protest against social injustice, to effect resource redistribution, and to generate empowerment (i.e. access to people, networks, organisations, skills and information). Since the image of places plays such an important role in attracting investment, stern anti-homeless and anti-squatter policies have been drafted (observable since the early 1990s). Even 'progressive' cities have distinguished themselves by adopting laws that prohibit people from sitting or lying on pavements in business districts.

Since such 'poor people's movements' usually have only their bodies and time as resources, their protest tends to be spontaneous and episodic, local in nature, and disruptive in strategy (and rarely led by major organisations). At best, their disruptive tactics briefly block normal city government operations and threaten local government legitimacy. In general, though, they face an increasingly recalcitrant and punitive state. Apparently only exceptional conditions allow them to unfold momentum and to gain success: such conditions need to be researched and specified further.

The new conditions on the labour market and the shift from social welfare to more punitive workfare policies have impacted on the urban movement scene in other ways. Not only have non-profit-making organisations 'run by and for the homeless' emerged, but the number and variety of institutions and projects 'servicing' the marginalised have exploded, many within municipal programmes harnessing the reforming energy of community-based groups. Their labour seeks not just to 'mend' the disintegration processes which traditional state activities cannot address, but they frequently develop 'innovative' strategies which already acknowledge the new divisions in society. Examples are grassroots organisations that help recent immigrants find jobs and places to live by training them to find work in the growing informal sector as day labourers rather than channelling them into normal job-training programmes, or projects working with unemployed youths through second labour market programmes training them as cooks or in housing rehabilitation (cf. Zukunft Bauen 1994: 12). Often the members of these projects are quite unaware that official politics increasingly looks to NGOs and community groups to replace state politics and to function as repair networks for economic and political disintegration processes, though the world summit in Copenhagen in March 1995 made this quite clear (cf. *Tageszeitung*, 10 March 1995).

Of course, the structural and contextual variables highlighted here do not by themselves explain the trajectory of particular movements; they do, however, condition it and define the parameters within which today's movement actors' strategic choices are more or less likely to succeed. Whether particular groups or movements frame the strains urban restructuring poses for their constituency in more transformative or more conservative terms depends on a host of intervening factors. Specific resource environments, the given universe of political discourse and specific local conditions and cultures are relevant for the particular unfolding of urban movements. But underlying and structuring those differences in ideological orientation, resource base, and strategies is a fundamentally new pattern of fragmentation distinguishing the urban social movements of the 1990s from those of the 1970s and early 1980s. The structural pre-conditions of movement activity have changed. Though there are apparent continuities within specific movements across this period, the latter were part of a broader and more united challenge to the broadening and deepening of domination and deprivation (Offe 1985), while the current urban social movements are far more fragmented and play a more contradictory role, not just because there is little overlap or resonance between different and more distinct movements, but also because these movements now reflect a new cleavage between 'insiders' and 'outsiders' and in part contribute to creating new exclusions. At the same time, the incorporation of many of these groups into the established political process has created new problems of complex governance: the new forms of regulation, which increasingly involve tripartite negotiation frameworks, have to be broad and flexible enough to process the complex antagonisms and varied systems of oppression which the political mechanisms of the Fordist capital–labour accord can no longer handle.

The implications of this social transformation for urban social movements are similar across the European landscape. Though the movements engage with locally specific issues in nationally diverging settings, these issues are not merely locally produced but are part of a larger transformation of global capital accumulation and state power. The competitive forms of urban development, the erosion of the welfare state and the expansion of the urban political system to include non-governmental stake-holders have been characteristic features of this transformation and have impacted on the development of urban social movements, fragmenting them in the ways described above.

Neither policies nor movements in the different European cities are, however, any more 'identical' than they have been in the past; local and national variation continues, possibly even intensifies, between the British, French or German experience and as multiple patterns of urbanisation emerge in the new hierarchy of cities. For example, we may expect specific social groups – among the young, the female, the old and 'non-citizens' as well as ethnic minorities – to be primarily affected by the new marginalisation processes, but they will likely be absorbed into and mobilised by movement milieux in larger rather than smaller cities. Metropolitan areas rather than medium-sized cities are the sites where (young, etc.) people's movements around social and labour disadvantage can be expected to widen and intensify. And whether new poor people's movements can emerge depends also on a variety of conducive factors, but first of all on the size of cities, because the transformation of unemployment and housing problems into actual conflict lines and the emergence of a critical mass are pre-conditions for movement milieux even to emerge.

In spite of such diversity in local incidence, the five kinds of movements which can currently be identified, in some form or other, across urban Europe, share certain features regarding their relation to an emerging model of regulation. Situating them in the broad context of urban restructuring helps to understand their contradictory role as both contributing to and challenging the new forms of regulation emerging on subnational levels. Thus, for example, their practice with innovative urban repair and their inclusion in the expanding urban political system can be seen as feeding into the search for new governance arrangements; their challenge of undemocratic and unecological urban development schemes may yet contribute to a more participatory and more sustainable First World urban development path (while preventing actual shifts of power). Some agendas would seem more effective than others, given the constraints of the context within which the movements have to operate today. But if we do not acknowledge and make transparent their new dependencies (on both state and market), it will be difficult to identify their possibilities.

Today, as so many hopes – including the hopes of municipal politicians – are placed on social movements and voluntary associations as agents of democratic revitalisation and for strengthening 'civil society', this task of identification seems especially unattractive. But the empirical evidence as well as analytical account of the tensions, contradictions, privileges and dependencies of the contemporary urban social movement sector should warn us against naively hopeful expectations that the power and destructive tendencies of the market and the state could easily be challenged by the heterogeneous groups making up this sector. Far from automatically fostering a 'communitarian' situation, this sector reproduces, within itself, the exclusionary and polarising tendencies characteristic of neoliberal politics. As we have seen, the movements themselves are affected, in various ways, by the (politically enforced) subordination of social life to market criteria.

Given the new structural conditions, their best bet might be to exploit the new arrangements of urban governance and the expanded boundaries of local politics, which have made new avenues available for those forces amongst the urban social movements that can seize them and that can tease out their ambivalences. Also, it would make sense to build on the mobilising potential of new inequalities and to politicise the social polarisation inherent in the 'global city' (i.e. the city that is subjected to globalisation pressures). The task of today's movements of the first kinds would be to turn their resources and their relative stability into support for those of the last kind, i.e. to attack and counter the exclusion and discrimination at the root of the new form of poor people's movements, and to exploit the dependency of the new negotiation frameworks on 'local residents' input' for the benefit of those marginalised and excluded.

Notes

1 In Frankfurt the SPD mayor even welcomed the first squats as 'symbolic actions': they served to legitimate and strengthen the local SPD's reformist efforts. After the third squat in November 1970, however, the mayor decided there were enough symbols and declared that further occupations would not be tolerated.
2 Obviously, research to draw on is more limited for this current phase than the last two. In particular, I cannot yet draw on detailed local case studies

for the unique formation of different cities' contemporary social movement scenes.
3 NIMBY means 'not in my back yard'.
4 For example, when directed against housing for asylum seekers. Explicitly right-wing and neo-Nazi movements and their attacks on immigrants and refugees, which can be described as another novel addition to the social movement sector, remain outside of this analysis, because it makes no sense to classify them as urban social movements.
5 One function of such mega-projects is supposed to be a socially integrative one, enhancing residents' identification with the city. Barcelona was obviously an example where the city was actually unified on the Olympics project. In other cities, top-down strategies of implementing a mega-project are rather manipulative of residents' interests and thus trigger opposition. Movements of this (second) kind as well as movements of the first kind are comparatively weak in strongly decentralised countries such as France: here, the governments of medium-sized cities have so far not yet become as dependent on the market as elsewhere and local programmes have sought to satisfy and include the majority of local interests.
6 The movement content in these programmes is relatively weak in countries such as Britain, where partnerships involving locally based voluntary groups and community organisations have been widely used in neoliberal regeneration programmes. But even here McArthur notes that the local activists are still 'more accustomed to protest and campaigning' and have difficulties with the 'bureaucratic procedures' prevalent in the community partnerships (McArthur, 1995: 67).
7 Amsterdam also has three sites of alternative housing, where people live in mobile homes, trailers, condemned buildings, vacant sheds and old buses, making up a variety of subcultures (cf. Deben, 1990).
8 The Senate of Berlin decided in 1991 not to permit any more encampments within the central area, where high-value uses and government functions are planned to locate.
9 'What we share and what unites us is the way we live, our lifestyle – collectively, without hierarchies, unconventionally, and with little dough' (*Vogelfrai*, 1994: 9).
10 Cf. Stephan Natz, 'Furcht vor weiteren Räumungen. Bunter und friedlicher Protestzug von rund 700 Wagendorfbewohnern und Hausbesetzern,' *Berliner Zeitung*, 25 October 1993, p. 19. Over Easter 1996, a national 'Wagenburgen' convention and demonstration was held in Berlin in order to 'break the isolation and make contacts with other social groups that are similarly threatened by the restructuring of cities' (a participant quoted in Asmuth, 1996:3).
11 When the first inner-city site (Waldeburg am Engelbecken with thirty occupants) was cleared by about 900 police and driven out to Karow at the northern edge of the city, some of the occupants, together with supporters, went on a sixteen-day hunger strike and held a vigil in front of City Hall, accompanied by actions such as leafleting parliamentarians, visiting the mayor, symbolically occupying the site, etc. (*Tageszeitung*, 4 November 1993).

7 The construction of urban services models

Dominique Lorrain

The notion that urban government and institutions can co-ordinate the development of large metropolises is today questioned by a threefold argument concerning: (a) their capacity to run large and complex metropolitan areas; (b) the role of markets and firms; and in more general terms; (c) the very principle of the possibility for strategic action.

Nineteenth-century political thinking, which generated local government in every European country, developed against a background of the existence of a degree of unity between production, exchange and sociability. However, this core feature of a golden age in local government is being called into question by the rising mobility of trade and population and by the appearance of a space proper to each of these elements: spaces for work, consumption, leisure, family. In place of a one-time spatial and political totality, a differential territoriality has come about with a retroactive effect on urban politics. Firstly, too many *communes* and authorities are responsible for various problems. City government has become overcomplex and thereby entails operation costs – transaction costs, for economists – as well as less control over decision-making on the part of those holding office (who are hardly in control of the complicated links they have set up). The result is a loss of understanding for the population at large (Lorrain 1989; Le Galès 1995). Secondly, it is by no means certain that these questions are best treated at the level of *communes* whose unsuitableness is conspicuous when it comes to the management of urban networks and the large technical systems. These strike across territories. Experts talk of discrepancy between functional and institutional territory (Offner and Pumain 1996). For some reformers, especially those who uphold public choice theory,[1] problems would be better resolved were they to be referred to the functionally competent institution. Such thinking has had a considerable influence on solutions adopted in Britain since the 1970s and the Radcliff–Maud Report. Elsewhere the operational soundness of small-scale local government is being questioned (Lefevre 1992).

The second factor questioning the validity of local government stems from the globalisation of the economy. The increasing role of firms and markets produces a concomitant reduction in the influence of governments and public actors. And local politics, rooted in its political and territorial legitimacy, is doubly ill at ease in this scenario: it is public and local. Moreover, between local government and the dominant internationalised firms there is an organic asymmetry, given the vast difference in financial, human and technological resources. This is an unprecedented situation for urban government and poses new questions – economic where the regulation of monopoly markets is concerned, but political too.

A third factor at issue here is the growing influence of the interactive paradigm in the social sciences. Our conceptions of collective action have moved in the space of a few years from a structural vision to an interactive representation. Nowadays, many disciplines give a central place to actors in the context of the action – for example, the methodological individualism of Boudon, the ethnomethodology of Garfinkel, the contractual agreement of the Conventionalists or the organised anarchy of March. All these studies place the actor at the centre of observation. One would be entitled to think that this would strengthen the notion of exercising command over collective action; in fact, the opposite result has come about. Actors perform but produce contingent solutions, end up with a purely local compromise. The change in emphasis between *L'acteur et le système* (Crozier and Friedberg 1977) and Friedberg's new book, sixteen years later (1993), exemplifies the path pursued.

At a risk of oversimplification, one could say that social sciences in the 1980s became more precise and more convincing in describing what actors do, at the same time as understanding of the global implication curiously diminished (Maurice 1994). By dint of insisting that everything is complex, contingent, interactive, localised and situated, the complete picture suffered an unintentional loss of perspective. Thereby the notion of an actor with a collective project being in a position to affect the system receded.

These three lines of argument have led to a disenchantment with the role of politics in general and of local politics in particular because, after considering that it was possible to exercise a significant influence on changes taking place in the world, the vision of collective action has become less of an aspiration. What place remains for political power in the governance of large metropolitan areas? Does the boat still have a pilot?

This dismissive perception of collective and political action is open to question. It is focused exclusively on the present moment, at which the complexity of the rules, technologies and actors involved removes the possibility for an actor to exert any significant influence on the course of action.

However, it is also worth considering that the subjects of collective action – a sector-based policy or a technical system – follow the cycle of Vernon products – birth, growth, maturity, sclerosis, crisis and redevelopment. The perspective at the point of maturity ought not to mislead us. If one adopts the stance of 'infancy' when new policies are drawn up, perception of the political role becomes markedly different. The asymmetry between political and industrial power is less absolute. Today's huge organisations were not always so huge. When they originated, they had less power and their room for manoeuvre was more indeterminate. None held a monopoly of know-how, because knowledge was still to a large extent indeterminate. A large proportion of the rules for action still had to be written.

In other words, if one reconsiders the role of politics and institutions by situating it during a market's infancy, or before it is set up, the possibility of an actor's exerting a strategic influence on the overall layout makes perfect sense. At such a time, urban politics has a notable role to play. It so happens that the history of urban networks provides us with a subject that is over a hundred years old, that is rich and complex and directly linked to the problem of urban government.

In the nineteenth century, when water services were developing and the first gas, then electricity, networks were being set up, there was the need to invent stable forms of collective action that we term models of urban services – others speak of an 'institutional regime' (Lorrain 1993). The term refers to the existence of a distribution of actors and procedures forming a fairly large body over a fairly long period. These models may be depicted on four levels:

1. The first level is from an organisational perspective or from that of the division of tasks between actors.
2. There is a further need for rules, legal procedures, pricing methods, a chart of accounts, and procedures for dealing with conflicts of interest.
3. Then, for one to be able to talk of a stable and reasonably durable system, it requires that the actors concerned share certain values. These questions are more central than might appear, even if combined attention is frequently centred on the actors. When all is well, the edifice appears quite natural; the actors fully involved in the action do not realise that certain commonly held values are the product of agreements that are in no sense automatic. The principles

governing pricing, equilibrium and subsidy perform of their own accord. In this phase only the forms of organisations are visible. They crystallise attention and reform.
4. Models of urban services develop from technical systems; thus stabilised technologies are needed for the whole to mesh.

How are these stable forms of collective action built up? In what follows, we shall present the stages in the process of construction with reference to the history of the gas and water utilities. This approach has the added advantage of calling upon available historical research so as to give due attention to the four phases in developing a 'model': technical competition and stabilisation; the development of an institutional framework; the effort to elaborate principles and norms and to create needs; and the work of legitimating these operations. Finally we shall draw the relevant lessons for present-day local government.

With such an insight into the role of political influence in collective action we want to stress three major arguments. First is the importance of technologies and techniques as part of modern life. Cities in the contemporary world can be described as large technical systems. And there are not two separate worlds, one of politics and general principles and the other sphere of technologies. Both are part of the same world; they are permanently interconnected. Technical choices in this matter deal with economics – its principle of efficiency and its logic of optimising productive needs – and with politics and its principles of equity and social justice. Second is the role of public actors, even if technologies have their own logic, and even considering the growing influence of large firms. The political sphere (*le politique*) has a role to play. It has to set up the rules of action; it has the responsibility to establish an institutional framework. To employ a well-known formula from institutional economics, the choice of constraints precedes choice in constraints. Third is the necessity to adopt a long-term approach to understand the dialectic role of markets, firms and political actors. What is said today to be a model – coherent and stable – is in fact the result of a long-term collective action, where private firms, political decisions, social movements, technical regulations and crisis have played a role. Starting from the origins is a method to describe the different steps and elements which have contributed to building a model. The way the problems were resolved a hundred years ago for gas and water must be of use in understanding questions of strategy today for telecommunications or electricity.

Consolidating the technologies

According to the account of Joel Tarr,[2] the historian of urban technology, the nineteenth century was characterised by separate spatial, city-based networks, which were not interconnected. The first gas company to provide a public service was set up in Baltimore in 1816. The gas was distributed by means of a network based on the architecture of water networks. It was a period when technologies, industrial systems and institutional models competed with one another.[3] Tarr makes a particular study of Pittsburg, where three family firms, concerned with producing pipelines, the work of distribution, and the production of electricity, were in competition. This initial socio-technical pattern of organisation lasted, through ups and downs, for nearly 120 years. Three main phases stand out – coal gas, water gas and coke gas. Then in the 1950s innovations occurred, utilities increasingly used natural gas and a new model organised the gas industry.

The predominance of coal gas

Before 1815, engineers avoided coal because of its impurities. The production of gas made use of different processes, involving pine resin, colophony[4] and wood from Carolina pine barrens. But engineers in Britain perfected new processes which could be employed with bituminous coal from Virginia and with imported coal. At the time the use of gas remained limited, and several factors restricted development:

- Transport costs bore heavily on the price of coal. A breakthrough occurred with the improvement of river navigation and, in particular, railway construction (Pittsburg was linked with Philadelphia in 1851 (Chandler 1977)).
- Gas transportation developed in line with the piping industry. Initially the industry used wooden pipes; then in 1820 cast-iron piping was developed in England and produced ten years later in the United States. Thus transport and storage became more dependable.
- The technology involved in piping affected the manner in which gas was distributed: hence its application. The quest for economies of investment induced the industry to lay small pipes which proved to be inadequate to meet an increase in demand.
- Initially, gas was used for street lighting but technical innovation in burners and meters opened up the possibilities for

domestic use, though high prices meant that it was only within the means of the better off. Oil lamps remained the main provision for domestic lighting.

These years were the time of the industry's youth, when the technology was not yet secure, application was still problematic and no major actor had yet emerged. Companies initially set up to provide public lighting, in the hope of making a profit, were obliged to lower their sights. The strict regulation applied by municipalities imposed limits on them, and the technological impediments drawn attention to above prevented them from moving out of the secondary market of public lighting and acquiring the consumer household market. In spite of this resistance, the gas industry developed. Between 1850 and 1860, coal gas networks increased from 30 to 221; by 1870 they numbered 390 (Tarr 1995: 7).

Water gas

Between 1875 and 1927, the gas industry was transformed by competition from other forms of energy. The first of these was kerosene, a by-product of oil. The more serious competition from electricity began in the 1880s when the companies concerned launched arc lamps onto a market which at the time was controlled by gas.[5] The effect was to drive the gas industry towards new technology and an improved and more viable service.

The first technical innovation concerned carburetted gas, a process consisting of passing 'steam upon incandescent carbon, usually anthracite coal or coke' (Tarr 1995: 9); the reaction produced a gas made up of hydrogen and carbon monoxide; the gas had a low calorific capacity (300 Btu) and generated a blue flame. It was enriched by the process involving liquid hydrocarbons which was patented by Lowe in 1875, and this process was acquired and developed by a gas company. The enrichment of water gas by using oil derivatives produced a gas whose performance in lighting was equal to that of coal but at a lower cost.

A further variant consisted of using oil rather than coal as a source of gas production. The first process was applied in 1899, and the first oil gas plants were developed in California.[6] Joel Tarr mentions that the inventor of this process was none other than the son of the Lowe who had invented water gas enrichment. In both instances, water gas presented advantages over coal gas. Installations were more compact and the gas had the thermic properties of greater flexibility. Gasworks using coal supplemented their means of gas production with equipment for oil or water gas.

Competition between the sources of energy also developed in the commercial field. Electricity companies attracted new customers by making a direct approach to industry. Gas companies followed suit in order to move into the domestic cooking and heating market. The policy varied between cities and the business aggressiveness shown by companies. Innovations, like the Bunsen burner, further extended the market. After 1900 product development turned more to industry; but the standards which applied imposed a limit. The thermal standard was defined as a quantity of energy per volume (450/550 Btu per cubic foot), but it was frequently in excess of industrial needs and hence imposed unnecessary expenditures. The gas industry lobbied for another standard, based this time on a thermal unit – the therm (100,000 Btu). It was thus in a position to supply customers according to their needs. Between 1919 and 1927, the use of gas in industry rose from 70.4 billion to 136.4 billion cubic feet. Competition with electricity continued, with gas companies making inroads into the sector by acquiring firms either manufacturing equipment or producing electricity. In 1899, according to Tarr, 40 per cent of the gas industry had interests in electric lighting; conversely, the electricity industry itself was moving into gas.

Coke gas

In this competition between technologies, coal gas did not entirely lose out. Its by-products gave it an advantage in relation to water gas. Coke obtained from coal could command substantial profits on resale.[7] At the time of the first world war, a revolution in the manufacture of cast iron from coke offered a first outlet. Since coke-oven gas had a thermal capacity this made it suitable for different applications and created a second outlet. In no time it proved a success. Cokeries developed in order to supply the various secondary markets – coke for smelting, coke gas for domestic use, and coke for producing water gases. In 1932, cokeries accounted for 25 per cent of the gas produced, which was then sold to public utility companies. Conversely, these companies developed their own manufacture of coke, thus enabling them either to cover surplus demands for gas or to meet a specific demand for coke. At the time, gas and coke industries were close to one another, with the manufacturing process shared and with compatible equipment.

A new socio-technical system

Following the second world war, the then existing socio-technical system finally gave way to a new source of energy: natural gas. Exit gasometers and gaslamps. Natural gas, in fact, had been known as a product for a long while. According to Tarr, the first discoveries of natural gas took place in Pennsylvania around 1820, followed by others in New York State. The first natural gas company in the USA was formed in 1858. When oil was discovered in 1859, natural gas followed as a matter of course as a by-product. The first transport pipeline dates from 1872 but it was only five miles long. Piping technology for a long time proved a hard nut to crack: there were difficulties in making pipes airtight and this restricted supply to towns situated close to gas-fields. Early in the twentieth century, improvements in pipeline manufacture made it possible to transport natural gas over long distances, thus new gas-fields could be exploited. This development took place in the interwar period. Between 1925 and 1935 a thousand-mile network linked certain western cities to the Californian deposits. This was constructed by big companies who had raised capital through share issues and new gas transport companies. The expansion would continue after the war. In 1944 the pipeline between the Gulf of Mexico and Virginia was completed.

The entire organisation of the sector and related practices was transformed. Local companies ceased producing and concentrated on distribution. 'They thus became spatially connected technical systems and part of a much larger nation wide network' (Tarr, 1995: 2). Coal gas which had been in existence for almost 150 years disappeared almost completely and natural gas became generalised. The transition to natural gas led to a complete change in general principles, in the type of regulation and in practices.

General principles Tarr and Hughes advance the notion of development by reference to a leading industry. Depending on the period, the gas industry has fluctuated between two basic systems – one resembling water distribution when the system was urban, and one resembling the electrical industry when the system became interconnected and complementarity in application brought the two industries together.

Regulation A local public service represented the first type of organisation, requiring municipal authorisation for street lighting and price fixing. Technically the system was little different from that for the supply of drinking water – a grid-based system, according to Kaijser. Gas which once travelled short distances now came from far away, crossing state frontiers, hence was subject to federal regulation. The territorial constituent of this socio-technical system thus affected its mode of regulation.

Practices The conversion to natural gas raised a transitional technological problem in principle similar to the one that occurred when the electrical industry moved from continuous to alternating current.[8] Conversion involved adapting the distribution system, changing household installations and a change of practices with the generalisation of the use of gas for domestic heating and cooking.

This brief summary of an industry's history over 120 years makes it clear that the turning point comes when new technology establishes itself. Until that point is reached, the system is impaired and suboptimal, and number of factors concur to limit its development. Several technologies are in competition, with none in a position to exercise a decisive advantage. Conversely, in order for a new system to be able to impose itself, it would require more than the discovery of new natural gas-fields; it would need the transformation of the entire socio-technical system and public authorities would have an important role to play.

What snags did the first socio-technical system come up against between 1815 and 1950? The exaggerated cost of the transport of raw materials; insufficient outlets for the by-products of the manufacture of gas because of an undeveloped chemical industry; a transport and storage industry that was at the mercy of pipeline technology; and the problem of diversifying beyond public lighting, which would have allowed economies of scale. Decisive change required a degree of transformation in each element of the socio-technical system; and in each instance municipal government had a task to fulfil: safety regulations for transport and storage, establishment of norms for domestic equipment, and so on. The techniques that we have referred to are meaningless without being associated with use and practice, and this in most cases requires regulations concerning hygiene and a degree of normalisation. Then, in the background, those who operate this energy sector remain local public utilities companies, regulated by the municipal authority. It goes without saying that discreet and prescriptive action of this kind scarcely conforms to what one normally associates with politics; nonetheless during this phase political institutions were able in many cases to play a role.

Establishing an institutional architecture

This capital and, in general, highly visible phase in constructing a socio-technical system presupposes the sharing of responsibility between institutions and organisations, which implies answering a number of basic questions. Who has responsibility for a sphere and the legitimate power to organise it? Who are the operators? How is the whole sector regulated? At what territorial level do these institutions function? These are primary questions with consequences that are of great importance for the history of the technical system and for that of the local authorities (Lorrain 1993). The history of industrialised countries in fact teaches us that choices made show considerable variation from one country to another. In one and the same sector, wherein differences associated with the technological variables are eliminated, international comparisons reveal that organising authorities may be municipal, regional or national. In some countries operators are public concerns, in others private concerns listed on the stock exchange. Regulation of such bodies may be undertaken by central government, or by regulating agencies, or else left to mechanisms of self-regulation.

These distinctions bear the mark of politics and are evidence of real differences in attitude, in political culture and in the methods adopted for dealing with conflicts of interest and for implementing categories of trust or mistrust. The course taken in elaborating these major institutional systems has been twofold – a gradual adaptation or a *moment fondateur*, i.e. a decisive starting moment.

The provision of water in France certainly belongs to the first category. *Communes* became *de facto* organising authorities because water is a local resource and the establishment of a network implies municipal authorisation. Private companies which are now so important were never specifically designated at a *moment fondateur*. They rested their case on the legality of the Allarde Decree of 1791, which asserted the liberty of commerce and industry and subsequently gave *communes* a free choice in fixing the term of operation.

Similarly with the development of the railways. In several countries they began as a result of local initiative. The Australian historian Salsbury explains how at the start separate lines were built by different companies to whom the notion of a comprehensive network did not occur. 'The need for a standard gauge, and for a uniform braking and signalling system, only became apparent when the first lines were laid and working' (Salsbury 1995: 32). And he introduces the term 'hazardous technical network'. Nevertheless, if political institutions are conspicuous by their absence in the early days of this first genealogy of networks, they soon make their appearance in order to lay down norms and safety regulations and impose uniformity in track-laying.

Quite different is the role played by political institutions in the path of change that we have called the *moment fondateur*. To this category belongs the reorganisation of gas and electricity in France at the moment of nationalisation (Caron and Cardot 1991), or the ongoing policies of deregulation (Lorrain and Stoker 1997; Bishop, Kay and Mayer 1994).

Let us return for a moment to the case of gas and turn our attention from America to the Netherlands where we have a textbook instance recounted by Arne Kaijser, the historian of urban technologies. It has an interesting story beginning with the discovery of the huge Groningen natural gas-field in July 1959 (Kaijser 1995). The transfer to natural gas, hence to a large interconnected network, implied the elaboration of a new institutional model, because the organisation could not simply depend on local authorities awarding franchises to private firms. The organisation of the gas industry in the Netherlands was not equal to the task of exploiting the discovery, and so it was a case of dividing responsibility between institutions, framing a pricing system in a new national context and identifying different potential applications, and central government had a highly significant role to play.

This case makes it clear how 'government and the two oil companies – Shell and Esso – were able to reach a solution in this complex issue in a remarkably short time' (Kaijser 1995: 1). Two and a half years after the discovery, a new institutional system was adopted allowing a division of responsibility. A plan for interconnection was elaborated, providing for the Groningen field to be linked to the other regional and national networks. A strategy providing for the sale of the new product in its several applications was developed. At this *moment fondateur* political actors had a direct impact on the organisation of the new model. But to appreciate what was new to the situation, we must be reminded of the past.

As in the other cases mentioned, the first gas networks began by being private, constructed for industrial firms whose needs merited investing in a gasworks. The sector's organisation evolved rapidly towards a 'municipal model'. At the start, cities intervened as regulators by awarding franchises and as customers for public street lighting. However, when they saw how profitable the activity was, they engaged directly in gas production. The first experience of this was in 1848, other cities immediately following suit. By the end of the century, 80 per cent of networks were municipal, almost 100 per cent fifty years later. 'Gas supply was seen as a public service that ought to be carried out by local authorities under monopoly conditions' (Kaijser 1995: 3). In the 1920s there began a process of regional interconnection between certain local networks, but since this regionalisation involved different technological processes, it was not easy to achieve a greater degree of integration.

At the time, plans for interconnection at national level were being studied, following the example provided by the Ruhr. Various commissions sat after the war, all of which drew attention to the economies of scale and the safety that would come about as a result. However, no change took place and the system remained regionally based. The interests and characteristics of the different gas companies were too distinct for there to be common ground for agreement. It needed outside forces, a dynamic to break the logjam; and this was precisely what the 1959 discovery achieved.

Transition to a new integrated national network required that two questions be dealt with. New organisations first had to be set up and game rules and a general frame for regulation defined. The process was accomplished in two stages: a period of 'secret' negotiations between the firms concerned and a few politicians; this took place over a year and was followed by a parliamentary debate. The outcome was that several firms were set up for the production and transport of gas. The skilful balance they achieved conveys the government's determination to keep control over this important source of fiscal revenues – the basis of the Dutch welfare state – and reflects the position of the two oil companies implicated in the discovery, who wished to be associated with them.

The actors further had to develop a strategy concerning the uses of so abundant a source of energy. At the time, three distinct markets were considered: the domestic market, principally cooking (price above 30 cents per cubic metre); the domestic heating market then guaranteed by coal (price between 10 and 15 cents per cubic metre); and the large-scale consumer market (industry and power plants). The general opinion was that the swiftest and surest way was to concentrate on the third submarket. However, the proposed solution forcefully endorsed by the Esso and Shell team was to aim at the domestic heating market. 'A household converting to gas heating would increase its gas consumption almost ten times' (Kaijser 1995: 10). This option took into account the possibility of using the network in place; but this would require the construction of a new high-pressure network enabling gas to be transported throughout the country and interconnection with regional networks. Further, consumer installations would need to be converted. Finally, of course, households would need to be won over to the idea of exchanging coal-fired heating for gas heating.

The introduction of natural gas was effected more rapidly than foreseen. Before the end of the 1960s, the new high-pressure network linking the entire country was in place and the whole conversion programme was complete. In under six years an integrated system for gas distribution had been carried through, and the commercial strategy adopted proved successful: 'more than 60 per cent of households connected to the gas grid converted to gas heating' (Kaijser 1995: 16).

Creating needs, establishing internal principles

Not everything can be boiled down to overall structure or to the division into who does what. There still needs to be agreement about how things are done. The study of urban services models as well as ongoing experiences in other countries at the present time[9] clearly emphasise that positive results depend on a series of 'preliminary conditions' being met. Does there exist a definition of public service or universal service within the general body of law? What contractual procedures are there to enable actors to formalise their agreement? How is the relationship monitored and regulated? And what is done in the event of a conflict of interest? How is the system financed?

All this appears settled and natural when the system functions well and the major technical choices (techniques, as well as laws and norms) slip into place as part of a background we are used to, whereas they too are the products of collective action. In order to show that actors have a determining role to play in the construction of laws, rules and norms, we shall develop the history of water distribution in France between 1850 and 1940. If one looks at the question from the viewpoint of market-building, the strategic role of big private firms is very much to the fore; they bring direct influence to bear on assembling a framework that favours them. But if one takes account of the need to create more farreaching conditions – a public space occupied by a local public service – then political actors come into their own. Let us take a look in turn at these two components.

The history of Pont-à-Mousson, the leading manufacturer of cast-iron piping in France, now merged with Saint-Gobain, is an 'interesting history because it shows how opinion is formed and develops, and how an industry directs it and harnesses it' (Baudant 1980: 229); and we shall further show how a market is built on the policies of a few large firms. The story is told in the history thesis compiled by Alain Baudant who made a detailed study of the firm's records covering the period 1918–39. He examined the accounts and scrutinised internal reports and correspondence between members of the board. Pont-à-Mousson was in a position to establish itself as the leading supplier of piping in the national market and the single producer of cast-iron. The analysis is supplemented in some particulars by the memoirs of the firm's chairman in the 1960s (Martin 1984). Baudant also shows how Pont-à-Mousson, though not a major service operator,[10] became an interested party in the history of urban services. Sales of piping indeed required the development of drinking water networks, hence a policy was to be promoted to this end.

The story begins in the aftermath of the first world war. The generalised laying on of water was held up by both cultural and financial obstacles, but the group confronted these in a strategy which met with considerable success.

The cultural obstacle

Towards 1925, 'the development of the market for main-pipes in France encountered the traditional resistance of the French to questions of hygiene' (Baudant 1980: 220). A number of studies are cited by the author in support of this thesis – Guy Thuillier on the Nivernais region; the reports of the Rockefeller mission which covered north-eastern France in the years 1918 to 1922; the treatise on hygiene (a classic) by Macé and Imbeaux, published in 1906. Baudant's findings are corroborated by other studies, notably by Eugen Weber (1983), who stops at the first world war. An indifference to hygiene was the behavioural characteristic of the French at the time in question.

Since the installation of piped water advanced slowly – too slowly for the group's management – a new policy was pursued, inspired by the example of Michelin, whose factories were visited by one of the Pont-à-Mousson board in 1927. As Baudant reports (1980: 227), 'it was not a case of declaring cast-iron piping to be the best piping in the world, but of launching a movement in favour of hygiene and water, of creating needs and customers'. And the group played a highly active role in the widespread promotion of water as an agent of hygiene in line with the message of late nineteenth-century reformers who propagated sanitary principles and practices. In 1927, Propex was set up, with the task of co-ordinating propaganda (now we should talk about a communications agency). But Propex adopted a clumsy stance on these delicate questions that were situated somewhere between the common good and industrial strategy. The firm's records recount a number of incidents that arose when executives from Propex, who were too closely linked to the manufacture of piping, descended on small towns and villages in France with the aim of selling hygiene and happiness to a recalcitrant flock. In October 1928 visiting speakers were received with bad grace by the mayor at Pont-de-Vaux (Ain), himself a doctor. Quoting from reports made by personnel from Propex, Baudant vividly conveys what took place. 'The mayor prevaricated and enquired of us whom we represented. He said that industrialists who sought to engage in advertising should at least pay for the hire of the hall. In some discomfort, we replied that we came with the support of the Ministry of Health and the Employers' Federation in the production of cast-iron piping.' And Baudant (1980: 230) stresses the difficulty faced by the personnel from Propex when he adds: 'all the while villagers hung around the vehicles with their Pont-à-Mousson sign'. Commenting on the incident, Roger Martin, who headed the group around 1970, remarked laconically that a choice had to be made between defending the cast-iron pipe and, more importantly, the issue of health (Martin 1984: 114). Pont-à-Mousson was obliged to keep a low profile:. Propex's credibility depended on this, and its structure was remodelled as the General Association of Hygienists and Municipal Technicians (AGHTM). Having abandoned any organic link with commercial interests, it was free effectively to discharge its mission for the common good. Trips were organised from October 1928 onwards in Bresse (north-east of Lyons), the Haute-Saône and the Dordogne, then in other regions. In retrospect, events show that the two factors which enabled the provision of a water supply in rural areas to be extended after 1928 were (a) the award of subsidies (as in the Meuse and the Ardennes on the grounds of post-war reconstruction),[11] and (b) an appropriate propaganda campaign (as in the Jura and Savoy, where the initiative of enlightened parliamentary deputies counted for a lot).

In their activities, speakers relied on contacts with administrative and political decision-makers, but also spread the word via primary schoolteachers and the 'general secretary' (top civil servant) in rural *communes*. A film would be shown and a discussion followed. As Baudant notes, the elements contributing to the success of Propex were 'the deputy, the film material – both comic and serious, the gramophone and the prospectus. Propex was a source of education. And of entertainment. And of pleasure, because it knew how to enlist people's sympathy for an event which was both educational and free' (Baudant 1980: 231). And thus an opinion was formed, was transmitted by the press and spread through town and countryside. The second element in the strategy consisted in making sure that it took effect by seeing that the financing of a water supply was made available.

The financial obstacle

Again, a brief historical summary is needed in order to appreciate the context and the impact of what took place. In the period between 1850 and 1920, water distribution as a policy consisted in keeping public drinking fountains and fire hydrants supplied. The public utility was a limited one. Naturally the service was free and was financed entirely by the municipal surplus of the operating budget, with investment remaining low. Hence the upgrading of installations as well as the development of new water networks represented an entirely new financial problem. How to set up a financial formula whereby local authorities remained solvent or were able to engage in public works without requiring of their electors that they carry the cost? Thus, in the early phase, lobbying central government was crucial because it produced most of the financing required.

Lobbying was effected through the Comité hygiène et eau set up in 1928. It had links with the piping industry, both cast-iron and steel, and with the world of politics through the columns of *Le Figaro* and *Le Temps*. In this way pressure was put on parliament to provide subsidies. The result was the creation in 1931 of a Caisse de crédit aux départements et aux communes, which had financial independence and responsibility for ensuring in part a system of loans to local government. Investment programmes followed, such as the nation-wide plan covering equipment (1928–33) and major construction schemes (1934–37).

But the strategy had a serious disadvantage. The financial package made over for an improved water supply depended on the general course of government policy. For all their efforts, the lobbying committee was unable to withstand the reduction of grants in a deflationary period. Policy for the water sector was sacrificed on the altar of budgetary policy: 'A decreased demand for loans from government eases a burden on the financial market' (Baudant 1980: 250).[12] Hence for the manufacturers of piping industrial activity was partially dependent of the fortunes of national politics and policies.

Thereupon a new strategy emerged which led to the financial system in force today. The exchanges reported by Baudant drawn from internal correspondence at Pont-à-Mousson reveal a clear analysis of the situation on the part of board members: it was up to local authorities to finance their own investment; they must find a new source of income, hence *sell water*,[13] in cases where supply was free or raise the price if it was cheap; in municipalities a system of concessions was favoured.

Baudant quotes from the minutes of meetings between the Minister of Finance, Paul Reynaud, the managing directors of Pont-à-Mousson, Tubes et Aciers, and La Lyonnaise des Eaux. There is reference to 'the acknowledged shortfall in the municipal supply of drinking water, the problem being that the price of water is too cheap and there is terrible wastage'. The chairman of Pont-à-Mousson proposes that 'companies which make it their business to raise capital on reasonable terms should be allowed to undertake further construction of the water supply system and increase the sale of water. Hence the price of water must be increased and cheap loans raised' (Baudant 1980: 250). A spokesman for the ministry declares that 'the Government is ready to participate in a publicity drive to demonstrate that the price of water must cover its cost and, further, that for reasons of hygiene water consumption must increase. At the same time, it accepts that the main effort will devolve on concessionary companies' (Baudant 1980: 250). Distribution of drinking water in towns and cities was open to development. The initiative would no longer come from government but from municipalities, backed by their in-house services or private companies. With the revenue from water fees they undertook to extend the supply network. In Baudant's words: 'The policy of water distribution came of age in France.' Relations between government, municipalities and industry were put on a new footing. State subsidies were reduced and loans raised by local government came to play a significant role at municipal level. The whole apparatus took longer to become operative in the countryside.

Gradually, the idea that water has its price took hold in French society, among elected representatives as well as among ordinary citizens. Prices were able to rise to a point where self-financing could take effect, thus covering a large part of the investment programme. Once there, the sector found a new organisational geometry. There were two main actors: the municipality and its concession-holders. There was financial independence, implying conformity with market conditions. For the manufacturers of piping, it meant a change of strategy towards lobbying, a change clearly envisaged by directors in 1939: 'This all suggests a policy change which will require us to review our commercial practice.'[14] They would now negotiate mainly with private concession-holders, which from a commercial point of view was not a disadvantage.

This overview makes it clear that something like eighty years, from 1850 to 1930, were needed to construct the legal and financial schemes to enable private operators subsequently to develop on a market basis. These schemes are still in force now and constitute one of the foundations of the economy of urban services in France. At the start of the period, the attitude to water was that it is a natural asset, free of charges and freely available. Anyone can sink a well, thereby reducing the need to be linked to a network; it will be noted that a similar attitude prevails in many cities in developing countries. Free access posed a problem for any independent development strategy. The unhurried measured policy pursued by firms consisted in constructing their market, by patiently pressing the needs of hygiene and domestic comfort, which eventually called for an industrial response, and by inventing a set of legal and financial instruments. It is a simple equation. One needs to start from the fact that water becomes an industrial asset. Hence it commands a price in relation to its volume as measured by a meter. A regulated scheme of pricing produces a cash-flow which gives firms their independence and enables them to devise an industrial strategy. The device of a price and a meter does not account for everything, but it is among the central factors explaining the fantastic development of urban service enterprises in France some decades later when urbanisation afforded them new markets.

Where do political institutions find a place in this chronicle largely enacted by industry and the strategy it adopted? Certainly one's first perception is of its having no more than a minor role, probably because the whole question of pricing and meters bore the stamp of so much common sense and the appearance of being so specialised, that at the time it was difficult to see it in context; besides, French political thinking on local government has never been very developed. However, one should be careful not to dismiss the influence of political institutions in this episode since, in parallel with the construction of a market, there was the need to create a public space. Stéphane Duroy's published thesis recounts the main stages (Duroy, 1996). In the mid nineteenth century, when the Compagnie Générale des Eaux was created and the first piped water supply developed, there was everything to be done. Firstly, the enterprise needed to qualify this activity: the boundaries of a public utility for water distribution needed a definition. This operation would determine the legal system to be applied – private law and magistrates' courts (*tribunaux d'instance*) or public law and administrative courts (*tribunaux administratifs et Conseil d'Etat*). At the beginning, the service was limited to the public drinking fountains and fire hydrants; a domestic water supply was considered to be a private matter. It needed many rulings from the *Conseil d'Etat* for 'the domestic supply of water to be recognised as a public service' (Duroy 1996: 331). The first of such rulings dates from 1877 and 1878 but, as is made clear in subsequent case law, a general atmosphere of uncertainty surrounded the question of what qualified as a domestic water supply in the early part of the twentieth century; and it was not until 1934 or so that changing attitudes made it clear that 'all occupants of houses should have the availability of a domestic water supply'.[15]

A further task consisted in establishing contracts. For a long time legal experts gave their attention to separating construction concessions from operating concessions, which implied there was a clear distinction between the constructional and operational phases. Until then they were merged; it might even be said that for a long time the construction market provided a partial source of concessions granted. The public authority then proposed that the construction industry paid itself for the cost of new equipment from operational revenue (Georgin 1931). It was not until the 1920s that 'the term public service concession was accepted in jurisprudence' (Duroy 1996: 36). Later, in the mid 1950s, leasing contracts made their appearance, thereby formalising a commitment by public authorities to ensure the financing of installations.

The evidence here is conclusive that, alongside the efforts of industry to construct a market, public actors – the *Conseil d'Etat* together with municipalities – involved in conflicts of interest played a part in the construction of the notion of public utility concerning water. The development of urban technical networks in France is a function of the two forces – public actors as well private firms – acting together.

Sharing values

A system becomes accepted when it is efficient and stable in its operation, but also when it is legitimated as a system. There can be no common enterprise unless the actors share the same values. The French urban services model has evolved in the way it has because the French and their elected representatives have judged it normal for a public service to be assured by private companies, for these companies to make money and for the service to be paid for directly according to its use; over the same period other countries have thought differently. In order to define such conceptions shared at any one time, we would speak of collective values or mentalities in line with Maurice Agulhon's use of the term (Agulhon 1988: 296). At first glance it might seem that these mentalities are the result of long gestation and the spontaneous consequence of action, but this would be to fail to see how such collective attitudes are built up.

This point about legitimised practice is all the more essential because we are concerned here with urban services. They form part of urban government, unlike ordinary industrial activities. They have to meet requirements of transparency, monitoring, fairness and justice. Such conditions are not naturally present, nor do they relate to properties which would be automatically included in the system. Quite the reverse, they result from action. They bear the mark of political values.

Pursuing the line of discourse adopted thus far, we will now recount the story of a scandal in the gas industry, which took place in Salford (Greater Manchester) at the end of the nineteenth century. The episode caused a considerable stir of national proportions, and finally contributed to some of the operational procedures applied in modern local government. But here are the facts such as John Garrard the British historian relates them (Garrard, 1992).

The affair began in February 1887 when the chairman of the Salford Gas Company (which was owned by the town council) sued a coal supplier for libel. The supplier in question had claimed that 'bribery, corruption and fraud had been rife in the municipality for many years', and that the chairman had done well out of it. Following a trial which received considerable notice in the local and national press, the industrialist was found not guilty. Shortly afterwards, the municipal gas board brought about the resignation of the chairman. An investigation was then set in motion and proceedings were started against the former chairman one year later. He pleaded guilty and was sentenced to five years' hard labour. Negotiations then began between the municipality and the same former gas company chairman – negotiations which naturally presupposed his co-operation – with a view to recovering the sums overcharged. In August 1889, on provision of £10,000, he was released on bail and disclosed the names of the firms that had bribed him. The municipality began proceedings; and among the names of those most implicated in the bribery and corruption was that of the coal supplier. New prosecutions followed with the municipality as plaintiff.

The case assumed national proportions. Other instances came to light. The coal supplier declared that he had practised bribery on a regular basis since 1879. The flood of disclosures produced two main effects: it revealed both the extent of corruption and, indirectly, the organisational limits of local government in a period of change. The debate on corruption served both to show up the faults of the system and to give birth to a new model.

What first emerged from this was the growing importance of the heads of municipal services, in particular those of the major technical services, who were individuals of high competence, very much involved in professional international networks and with close links in industry. At the close of the nineteenth century, municipal activities were becoming increasingly specialised. Decisions were being taken more and more by specialised committees rather than by the council in full session; and in such committees a small group of individuals played a leading role, which explains why controls had not functioned properly.

The reasons for the scandal and for the emotions it aroused as well as for the extraordinary degree of support shown by the councillors for their gas department chairman are also to be found in the weakness of political parties and in the poor organisational structure of municipal services. It all came down to personal relationships. Elected politicians did not exercise a full-time mandate. For many, their position represented a charge, in the primary sense of the term – a time-consuming burden. Hence they were inclined to delegate. The end result was that decisions were taken by a small group of men. Nor was there any functional distinction between elected representatives and those they employed. Those involved were few; interpersonal relationships were paramount. Many municipal councillors who by rights represented the 'organising authority' looked upon the chairman of the gas company as a friend. Furthermore, at the time, the chief administrative heads were as well known to the population as the elected representatives, with the result that in addition to their specialised qualifications they possessed some small popular legitimacy.

These goings-on must also be seen in the context of the time. Commissions were common practice in gas and other services. The controversy that ensued in the professional urban services journals – *Journal of Gas Lighting, Journal of Water Supply and Sanitary Improvement, Gas and Water Review* – throw light on the practices and reveal them as widespread throughout the textile industry, as well as in the building and property sectors. In Garrard's words (1992: 181), 'all this suggests that those who headed the gas companies imported into their profession and into the municipal function habits and attitudes that were already current in the larger commercial sectors'.

This farreaching affair further highlighted an ill-defined boundary between administrative and industrial activities. The categories of the public service were incomplete in their construction. Public action at a local level was beginning to adopt formalised organisational procedures. In their infancy, professional networks of local administrators had not yet developed their own value systems. The one group clearly to emerge were the network engineers, but their professionalisation was still incomplete. In some respects they belonged to the local administration, in others they were part of the world of industry. This ambivalence in individual demeanour only reflected the uncertain status of the gas company: part public utility, part industry.

Put differently, the Salford gas scandal is evidence of the need to have clear, publicised collective rules and procedures that are fair and accepted; this is the definition of the making of a public space. At the time, a clear demarcation between a political function – elected representatives – and a productive function – the administration and the heads of network companies – was still lacking. The law governing public utilities as framed still fell short of distinguishing a public utility from an industrial activity. Gas found itself in the midst of these contradictions. The scandal helped bring about a change in collective representations. It played a part in the emergence of new rules for collective action at the close of the nineteenth century and it served to question contemporary cultural and ethical attitudes and also what passed as lawful.

Finally, the scandal has an exemplary quality inasmuch as it informs us about the ways in which group mentalities are shaped. It brings violence and passion back into focus. It establishes the role of crisis in the construction of collective rules. It teaches us that the work of politics and institutions is carried out at different levels and on different occasions. Patient and protracted labour is on occasion supplemented by turbulence: an exceptional and intense phase during which the actors shift the boundaries of their representation.

The terms of political activity

This lengthy detour involving municipal gas services and the domestic water supply has enabled us to pass through each level of a structure that we term a socio-technical system or urban service model, so as to indicate that it represents a stable and internally coherent whole. We have examined in turn the technologies, the choices of institutional framework, the focusing of principles governing action and the formation of collective values.

These developments can connect with a reflection on the nature of power. The power wielded by major firms would appear to be excessive, given the asymmetry between their means and those of local government. If one looks back to the origins, to the time when the models took shape, one is strongly inclined to question the idea that political influence has disappeared. Certainly the firms in question are powerful. Their strength comes to a large extent from their capacity to anticipate which leads them to act before the rest as system-builders (Chandler 1977; Hughes, 1983). But their structure remains fragile. The equilibrium achieved after many difficulties may be modified at each level of the edifice. And the organisations know that their power is never absolute. They are permanently obliged to protect themselves from rivals and justify the choices made to consumers/users. They must be legitimate and useful in order to continue to exist; the numbers of defunct firms which populate the scrapyard of 'good ideas' are there to remind us of the fragility of power.

And the result is to rehabilitate politics as a level in society. In its period of infancy, when a system is constructed, the game is wide open. The business of construction is not only due to the larger firms; it would appear to depend on the involvement of many actors whom we already have some idea of – entrepreneurs, engineers, legal experts, politicians and users' movements. In any event, politics will have played an important role, though abiding by certain conditions.

1. Development is over a long period. Gas technologies in the United States were in competition for more than 120 years. In France, in the water sector the system of delegated management began in the mid nineteenth century and the edifice was more or less in place before the second world war. This initial feature poses a challenge to local government, which too frequently follows its own pace and is ill equipped to exercise vigilance over such a time-scale, particularly as the process of development involves non-uniform progression. Periods of calm may be followed by periods of crisis and questioning, and both are instrumental to the construction of collective action. Political effectiveness is very often associated with intervening in the critical stages of a pattern's being redrawn; one thinks of the management of Groningen gas or the handling of scandals. But its effect is also felt across a string of minor interventions over aspects which at first glance appear insignificant. And action of a routine kind carries the more weight because rules and procedures cannot all be decided in one go. Hence the political actors must be able to adjust to two different time-scales: one involving the handling of crises, the other incremental action.
2. Intervention has no meaning unless it occurs at every level of the edifice: technologies, framing, principles and values. Each one is in its way important. For the politician there is no preferential level, no division of labour whereby he would be entrusted with the major institutional reforms, while to engineers and legal experts would be consigned the choice of norms, contracts and pricing. The politician's efficacy in the construction of collective action is dependent on his being everywhere.
3. This notion has some substantial bearing on the future role of public authority and provides a focus for a clearer definition of the function of government. Its first function – the oldest and best known – is that of *problem-solving*. The political activity of the state is legitimated because it produces goods and services. For a long while this conception was generally endorsed in Europe with the development of state-owned industry (Wright, 1993; Lorrain and Stoker, 1996), but deregulation has called it into question. What is there left for the politician if major prerogatives are gone? Second remains the function of *general steering*, and that of the overall organisation of responsibility, providing a reply to the question of who does what.

Within this role is the function – and a very important one – of devising the procedures for action and the instruments of control. Third, the sectors' histories we have recalled show that urban political actors are entitled to say what is legitimate or not, what is allowed or forbidden, and to create a space of legitimacy which may thereafter serve as a basis for the actors. Under this role political actors are *the guardian of ethics and norms*. As we have remarked, that part of the edifice which has to do with collective values and mentalities seems so natural that we take little notice of it. Nevertheless it is essential.

Today in telecommunications and in information technology a new view of public service is being formed. Traditional questions are re-acquiring actuality. Where does the boundary of this or that public service lie? Does it include access to services? At what level should the lowest rate be fixed? Problems are emerging which have to do with fundamental liberties, as for example that requiring a limit to be set on the use of information on individuals that is bound to come into the possession of large-scale operators – those concerned with pay-television, internet uses and so on. The public authority also intervenes in order to redefine older sectors that appeared stable – electricity, transport and soon, no doubt, water and refuse. In short, even if the economy of urban networks and large technical systems is in the hands of large private enterprises, the public authority is directly legitimated to intervene in the period of infancy when a frame for action needs to be 'invented' like when the system seizes up and requires relaunching.

Three functions stand out in a renewed role for urban political activity: an operational function, a steering function and that of guarding values. The balance between these three components may vary but there is a place for city governance so long as it manages different scales of time and adjusts its performance to each part of the model.

Notes

1 For a critical view see King (1987) or Dunleavy (1991).
2 Our discussion of the gas sector is based on the contributions of J. Tarr and A. Kaijser to a conference on Large Technical Systems, organised by a research group of CNRS (GDR réseaux, Autun 1995): Coutard (forthcoming, 1999).
3 The telecoms sector plays the same driving role now.
4 Colophony or rosin is a resin obtained when turpentine is prepared from dead pine wood.
5 A case of mimetism operating between sectors. We have already noticed that the gas industry took water as its model; gas in turn served as a model for electricity. 'The historian Harold Passer observes that Thomas Edison, who

The construction of urban services models

invented the incandescent electric lamp and built the first electricity generating station, had made a close study of the gas industry' (Tarr 1995: 9).
6 Europe, being without oil resources, relied very largely on coal.
7 Coke obtained in the water gas procedure was of lower quality; it was often used up by the companies and proved ineffective in generating by-products such as ammonia and phenols.
8 See Salsbury (1995).
9 Several cases have been documented and brought together in one volume: Caracas, H. Coing; Buenos Aires, D. Faudry; Djakarta, E. Baye; Sydney, J. Moss; Macao, D. Rétali; Italy, M. Venturini; the Ivory-Coast and Guinea, J. Cl. Lavigne. We have drawn up a number of rules for collective action in our conclusion (Lorrain, 1995a).
10 In the 1920s, the company diversified into water utilities with a company named Socea, which became Sobea in the 1970s, then Cise in the 1980s, and finally merged with Saur in the late 1990s (part of the Bouygues group).
11 The Meuse, Ardennes and Aisne along with other eastern departments were among the 'liberated regions' after the first world war and, as such, received a special grant from lottery money. Between 1923 and 1930, these departments recorded the highest mean piping mileage laid in France (Baudant 1980: 224). Furthermore, the Pont-à-Mousson group played an active role there on its own territory, being well acquainted with the problems and having close access to all the decision-making networks.
12 See Baudant's masterly study and the first subsidy crisis in 1934.
13 Our emphasis.
14 Note from Georges Morin to Marcel Paul (9 February 1939), in Baudant (1980: 250).
15 Circular dated 25 October 1934 issuing a general direction in respect of the provision of drinking water in communes, cited by Duroy (1996: 35).

8 Private-sector interests and urban governance

Patrick Le Galès

As argued in the introduction, one key dimension of cities is the extent to which various sorts of actors – more or less organised social interest groups – are brought together in processes of governance. If the approach put forward in this volume makes sense, it is necessary to look at the ways in which various interests exist, the extent to which they are organised within cities and how interests and institutions interact. This chapter is a limited contribution to the debate which focuses on private-sector actors and interests and the ways in which they partake in processes of urban governance.

It is generally argued that although cities and states were highly interdependent in Western Europe, and to some extent are becoming more and more so, many cities (in the sense of collective actors) have acquired an increasing role in political and economic terms (Le Galès and Harding 1996). The restructuring of nation-states and of the economy has created space for subnational mobilisation, especially at city level. Evidence of cities' economic strategies and of increasing political autonomy has been put forward in various comparative pieces of research in the past ten years (for instance, Judd and Parkinson 1990; Harding *et al.* 1994; Heinelt and Mayer 1992; Dunford and Kafkalas 1992; Le Galès 1993; Harding 1996). In most of the research in comparative urban politics, it has been argued that urban governance is moving towards serious changes in various European countries, the analysis of which requires a better understanding of interests in cities.

In this book, we have tried to analyse the environment in which urban interest groups operate, which has changed significantly in recent years. In her chapter, Margit Mayer notes a number of changes in the socio-economic environment which challenge the established 'urban social movement' approach to understanding current interest group activity in European cities. The most important conclusion she draws is that urban politics can no longer be defined essentially as the politics of collective consumption. Whilst this remains an important feature, there is also an urban politics of production which is associated with different forms of interest group activity and generates some complex alliances between state, market and 'movement'. Urban politics has never been dominated, nor has it ever been appropriately defined by, collective consumption issues alone. Urban production issues have recently increased in importance but they have not appeared out of nowhere. What Margit Mayer successfully demonstrates is the importance of placing changes in the salience of different urban issues in a much wider context. Interest group mobilisation cannot be seen as entirely voluntaristic. The patterns and forms it takes are largely made up of the economic, fiscal, statutory and regulatory environment in which it is rooted. The politics of production has grown in importance, while even in the sphere of collective consumption there is a much greater role for markets and those who can compete within them. Specifically, these changes have led to the mobilisation of groups that were not particularly active within urban politics in the past, but have recently become active, in development issues in particular. They have encouraged existing groups to mobilise in new ways and to respond to new themes and opportunities. For instance, the changes in the nature of their core activity have to a considerable extent forced groups in Great Britain to operate in a 'business-like' fashion in quasi-market environments, while keeping one eye on their own institutional survival. Such changes have spurred new, more strategic, more conciliatory and less conflictual modes of action.

Processes of European integration and globalisation define a path for changing forms of urban governance and reinforced patterns of co-operation and competition between European cities. Increasing economic competition and increased market exposure lead to forms of territorial competition. Private-sector actors play a more important role. They may engage in processes of coalition building and/or contribute to the reinforced fragmentation of some cities and their lack of capacity to structure local societies and resist/adapt to the globalisation trends. In any case, the hypothesis is that among the factors likely to play a role in explaining the making of an urban governance regime (or not) and its political orientation (more or less pro-competitive) in European cities, the type and organisation of private actors is among the most significant.

Due to the increasing mobilisation in favour of economic development and competition, attention has clearly shifted towards the limits of local government actors and the role of private sector individuals and collective actors. Examples of public–private partnerships in urban projects have spread rapidly in most European countries (Heinz 1993). Despite an apparent similarity to trends in American cities, this pattern has varied remarkably from country to country, and from one city to another.

This chapter briefly examines the extent to which economic interests are organised and involved in certain European cities. It does not consider any case in detail, but rather seeks to provide an overview of the ways in which business organisation and actors tend to be incorporated in urban governance processes. First it examines the type of actors (collective and individual) most likely to take part in urban politics, and then looks at the various kinds of interactions in which they engage with other actors, including local authorities. The argument presented here is that beyond the 'parasite' firms, which tend to take advantage of produced collective goods, and those who, for various reasons, are unable to leave the place they live in, some private-sector actors and/or organised business interests are likely to contribute to urban governance and orientate (at least in part) their strategies in order to contribute to the production of certain collective goods. This chapter, like the rest of the book, concentrates mainly on European medium-sized/regional cities, which make up the best part of the European urban structure.

Business interests and business actors in cities

In the current *intervalle historique*, which we identified earlier, urban politicians and private sector actors are tending to interact to define collective projects for the city. Private-sector interests are usually examined in terms of organised forms of 'old corporatism' (based on guilds and professions, most of them destroyed by nation-states and liberalism) and 'new corporatism' (trade unions, business organisations and the state), that have flourished in different ways in many parts of Europe. However, the literature of organised interests has shown that infranationally based corporatism has rarely replaced national corporatism. Private interests may not be so organised, but there are indications as to the mobilisation of private actors in different kinds in cities.

Organised interests? Employers and local Chambers of Commerce

European cities and towns in the Middle Ages were first and foremost cities of entrepreneurial merchants, where according to Black (1984: 237) 'the corporation organization of labour and liberal values developed simultaneously and in the same milieu from the twelfth to the seventeenth centuries ... Guild and "civil society" were distinctive features of European society.' Urban oligarchies included various mixes of bourgeois and aristocrats, of established families and new entrepreneurs. Historically, therefore, private interests and European cities were closely intertwined. Then 'voracious states' gained the upper hand over 'obstructing cities' (Tilly and Blockmans 1994) and cities progressively ceased to be the main locus for the structuring and aggregation of interests. Interest organisations became national, contributing to the strengthening of nation-states in Europe in the twentieth century (Crouch 1993). However, Crouch mentions that the organisation of interests in some countries (such as Germany and Austria) came before the rise of the nation-states and remained remarkably strong at city or regional level. The late formation of states and the absence of revolutions also allowed some interests in countries like Italy to remain powerful over a long period. In Florence, wealthy aristocratic families passed on their wealth and *palazzi* generation after generation (wealth which was sometimes first accumulated at the time of the Medicis).

Analysing interests in cities comparatively is an impossible task, as the research on the subject is scarce and uneven. National organised interests on the other hand have attracted a fair amount of research over the past twenty years, partly due to the debate on corporatism.[1] Similar research has not been carried out at urban level in a systematic way. Business organisations are no easier to deal with. Also, the organisation of business interests in capitalist societies and its territorial dimension is a delicate issue.

Claus Offe has identified the structural advantage of employers in capitalist societies. In a famous article, 'Two logics of collective action' (Offe and Wiesenthal 1980), it was argued first that capitalists were able to pursue their interest through the market and therefore did not need to organise, and second that their interest was less fragmented and diverse that that of trade unions. However, Streeck's empirical research (1992) in European countries has shown that the number of business organisations was greater than the number of trade unions and that there was considerable fragmentation and diversity between sectors and territories. At the very least, distinctions had to be made between employers' organisations, trade organisations and Chambers of Commerce and industry, but a great amount of differentiation prevailed in most countries. Often, business organisations only pursue a limited segment of the interest range of their members. The territorial organisation of interests has several dimensions. Streeck and Schmitter (1985) first suggested making a distinction between: 'the logic of membership [which] is governed by values and interest perceptions of the groups and individuals that an association undertakes to represent . . . [T]he "logic of influence" . . . consists of the constraints and opportunities offered to associations by their institutional environment, and it is experienced by associations as a set of strategic imperatives, rules of political prudence and norms of reciprocal exchange' (Streeck 1992: 105). Wolfgang

Streeck added to the complexity and suggested taking into account territory as a dimension of organisational structuring of interest distinguishing several dimensions (for the complete picture see Streeck 1992: 107–8):

- territory as a place for identity and identification of subgroups, and face to face interaction, i.e. territory as proximity;
- territorial subdivisions required to represent members in relation to subnational government.

Therefore, in cities, the comparative analysis of the political mobilisation of business interests is clearly problematic. Not much research has been done to date (see Peck and Tickell 1995; Waters 1995; Bennett and Krebs 1991), but fragmented elements have been used in various research projects. Regions have to some extent attracted more research in this area (Rhodes 1995) but the organisation of interests at regional level seems to remain quite weak except in federal states (Coleman and Jacek 1989; Trigilia 1991; Le Galès and Lequesne 1997). European integration has led to changes in the way organised interests are structured and operate at different levels, and that seems to be true in both economic-sector and voluntary-sector interests.[2]

Let us consider employers' organisations for instance. In his major comparative work, Crouch has shown the enduring characteristics of certain forms of business organisations (1993). In Germany and Austria, powerful local Chambers of Commerce pre-existed the formation of the modern nation-state. They were direct descendants of the medieval 'old corporatist' organisations. Even if German employers' organisations were powerfully structured at central level over a period of time, Chambers of Commerce have remained powerful bodies with expertise, resources and legitimacy. They have been involved in the process of strategy-building in various cities such as Hamburg (Dangshaft and Ossenbrügge 1990) and Stuttgart (Hoffman-Martinot 1995). In Scandinavian countries, the early centralisation of the state went hand in hand with the building of central employers' organisations that were rapidly involved in tripartite negotiations at the national level (Crouch 1993). In those relatively small centralised countries, local employers' organisations remain weak and do not often interfere with a robust welfarist municipal government. Until today welfarist Scandinavian countries have been characterised by strongly organised economic interests at national level.

In Italy, France, Britain, Ireland or Portugal, employers' organisations were never very strong, nor were they very active locally. In France and

Italy, Chambers of Commerce and Industry (CCI) are public bodies, but they have limited power to act as business representatives in an active sense, although this may vary from city to city. In Lyons or Lille, CCIs have exerted a more consistent influence due to their embeddedness in dynamic (often industrial) economic bases. In Britain, Chambers of Commerce and Industry were hardly more than business clubs, with some exceptions such as in Birmingham.

Things are not completely immobile and changes do take place, albeit slowly. In most cases in Scandinavian countries, even timid decentralisation reforms have not led to the serious empowerment either of Chambers of Commerce or local business organisations, with the possible exception of small cities in crisis, for instance in small paper-making towns on the Finnish coast. Powerfully entrenched economic interest organisations remain centrally organised. However, business organisation should not be considered as static. In Sweden, for instance, the employers' confederation, SAF, historically a powerful national institution, has argued against the government in order to promote decentralisation and local negotiation (i.e. at firm level) (Pontusson 1996). This does not automatically lead to the development of employer interest organisations in cities, but it may provide a more open arena within which some cities will be able to play an increasing role (as collective actors) and integrate some employers' organisations.

Local Chambers of Commerce and Industry and of craftsmen are another form of organised interest which requires examination. In contrast to employers' organisations, local Chambers tend, in any case, to have weak national organisations, so the local (here urban) level is the important dimension. Again, local Chambers are not a new phenomenon in Europe. In many European countries, guilds and corporations progressively turned into local Chambers. In some places, they gradually became residuary bodies; in others (Germany and Austria), they have remained remarkably strong. In France, Chambers date back to the early seventeenth century (Conquet 1976 quoted by Waters 1995). From the very beginning, Chambers were created at the French state's instigation to control private entrepreneurs, to regulate commercial and industrial activity and to obtain information. Abolished by the Revolution, local Chambers were re-established by Napoleon as part of the French administration to control the private sector. As consultative bodies, they mainly perform administrative functions on behalf of the state. Besides this, they mainly manage infrastructures, provide technical assistance to local firms and play a role in the provision of some specialised training. Increased urban economic development policies from the early 1980s onwards first provoked strong opposition from local Chambers. Building on their weak resources and weak legitimacy in most cities, most Chambers of Commerce and Industry engaged in a renovation process to face urban government leaders. They changed their organisation (becoming less oriented towards local notables), increased their expertise and services and often established themselves in a brand new building to symbolise the changes.

In Italy, local corporations lasted far longer than in France. The local Chamber model was more or less imported from France, first created in Florence in 1770 (Waters 1995) and many creations followed the Napoleonic invasions. The newly created Italian state enshrined them in detailed legislation, as did Mussolini when he transformed them into powerful provincial councils tightly woven into the fascist state. Although local Chambers were re-established, they were not reformed and remained under the strict control of the state or, put another way, became a locus for political party influence. They exert certain functions for the state and respond in varying degrees to local firms' needs. In the 'Third Italy', Chambers were praised for their role in institutionalising business networks (Nanetti 1988). The resilience and remarkable dynamism of the traditional economic sectors could make some local Chambers' traditional organisations represent and articulate interests. However, in most cities, Chambers were integrated within networks of political organisations (invariably Christian democrats until the early 1990s). Changes came through when the state withdrew its funding and local firms had to increase their contributions. As urban politics tend to become more prominent and legitimate in the aftermath of the direct election of a new generation of mayors, Chambers may gain greater prominence in some cities, for instance in Milan.

In Britain, the combined forces of liberalism and the industrial revolution did much to dissipate the old corporations. Liberalism was also instrumental in preventing the creation of local Chambers as public bodies, and of course neither Napoleon nor his model of public administration ever crossed the Channel. In most cities Chambers of Industry and Commerce were little more than additional clubs for business men, frequented and run mainly by shopkeepers. However, in some cities such as Birmingham, Manchester and Edinburgh, Chambers were able to play a more important role, and things were to change in the 1980s with the new national politics (see above).

If we set Germany aside, urban organised professional interests are to be found in most European cities, though they remain modest in size: employers' associations have been organised rather at national level (with some exceptions), and local Chambers are generally rather weak institutions within cities. Occasionally, some Chambers have played a significant role, thanks to a strong economic base and favourable political circumstances (in Birmingham, Barcelona, or Lyons, for instance). The changing environment in the 1980s did lead, to some extent, to organised urban professional interests gaining influence, but certainly not to the extent sometimes expected by corporatist writers.

However, instead of concentrating on organised business interests, individuals, firms and loose networks of private actors have become increasingly significant in European cities, either in their support for processes of urban governance, or in increasing fragmentation. Furthermore, the frontiers between public and private are tending to be blurred at least as much as at national level. For instance, public agencies working as private organisations, or private firms involved in community development, or the emergence of semi-public agencies, quangos and others make it difficult to clearly distinguish public and private organisations.

Private-sector actors, individual and collective

Private-sector actors in cities are weakly organised or not organised at all. For the sake of clarity, the following section attempts to list the various actors which may be taken into account in different cities.

The formal organisations representing businesses at local level – Chambers of Commerce – have only rarely been key players in the new urban politics. This is perhaps less surprising in the UK context than in countries where Chambers are much stronger (Bennett and Krebs 1991), but it does highlight the fact that new interest group relationships tend to work through personal networks rather than through established, bureaucratic channels. Case study evidence tends to suggest, for example, that when the leaders of city councils who were most associated with having changed approaches to economic development in their areas started building links with local business communities, they did not choose to work through the established institutions. Rather they tried to build more personal networks of people who were actually developing new projects within the city or showed a keen interest in doing so. The reason for this is simple: new relationships tend to be built upon concrete achievements. They therefore develop between individuals and organisations who have the capacity to act rather than the right to represent.

The private-sector leaders who show themselves keenest to act in urban politics are not necessarily those whose businesses are most dependent on the local economy. It is not the level of local dependency that matters, but the potential for improving business prospects that can stem from participation in various urban coalitions and networks. Quite how business prospects might be improved, and who the likely beneficiaries are, inevitably depends, to a significant extent, on the nature of the project(s) concerned. However, businesses are able to perceive business advantages in taking part in activities that do not necessarily lead directly to work for the company. On the one hand, they can provide opportunities to mix with others, in business or city government, who may generate future business simply on the basis of personal contact. On the other hand, they might offer opportunities to exert some influence on strategic choices which make it more, rather than less, likely that certain sorts of business that is beneficial to the company will be created. Lastly, opportunities for social prestige, political influence and patronage may also be part of business leaders' motivations.

It is too mechanistic, however, to assume that business leaders become involved in city affairs simply because they are asked to. Indeed, there is an argument that the involvement of business communities in urban politics through externally imposed institutions may ultimately prove to be one of the least effective models (see Peck and Tickell 1995). The UK has gone further along the institutional route to business involvement than most of its European partners. The ostensible advantages are that business leaders can add essential expertise to what are usually public-sector-driven agencies, or can at least facilitate a transfer of skills. They may also have alternative philosophies, methods and views that can widen the horizons of 'normal' public-sector companies. Quite whether they have fulfilled either role, however, is arguable.

The UK case is remarkable in this respect. It is possible to find examples in other European cities of business leaders who exert strong urban political influence, but probably mainly in cases where their firms enjoy a certain prominence. Thus the Agnelli family plays a dominant role in Turin.

Beyond the case of major business leaders brought together as nineteenth-century entrepreneurs were in Manchester and Birmingham, another factor is worth pointing out. The fragmentation of business organised interests makes it easy for either well-connected individuals or loose networks to play a role. Young entrepreneurs, for instance, often react to heavyweight traditional Chambers and tend to be active on a different basis. In some new economic sectors, loose associations are likely to emerge to support their case. Beyond that, private actors tend to assume different forms: professionals (on an individual or collective basis), consultants, small service firms, various kinds of shopkeepers. All these may have either an economic interest or some other interest in participating in the urban governance process.

Economic and political centralisation contribute to limiting the expansion of this sort of process on a large scale. In France, for instance, the modernising state led the way to economic restructuring and the concentration of most economic sectors as well as the demise of small firms. The training and recruitment of the national elite are also crucial. Bauer and Bertin-Mourot (1995) have demonstrated that most business leaders in France are former civil servants from *les grands corps*, particularly in industry and finance. These business leaders have a national view of the world and pay no attention at all to urban or regional issues, this being a job for state representatives. A not so different story could be told in Ireland or in Scandinavian countries. The structure of the banking system (so crucial in the case of America) is also quite telling. In traditional centralised countries such as France, Britain or Scandinavia, local or regional banks do not exist except in the co-operative or mutualist sector (in Brittany or Alsace in France). Germany and, to a lesser extent, Spain, Italy or Belgium have always had regional banks, or banks mainly located in cities likely to join coalitions. Mutual banks, *casse de risparmio* in Italy, have played a highly significant role beyond economic development in structuring networks and supporting certain projects. Again, economic and political centralisation usually structures the organisation of the banking sector.

The importance of building firms, utilities and private developers, reveals the weight of urban capitalism. In his analysis of the regional dimension of the organisation of economic interests, Van Waarden (1989) noticed that two sectors had systematically strong regional/local organisations: the food industry and building. Builders have always had an interest in urban politics, an interest only increased in the 1980s when national programmes were more or less in retreat and city leaders gained a greater say. The building sector in most countries was very fragmented with numerous small and medium-sized firms. However, over the last two decades or so, the construction sector – like others – has seen massive concentration and reorganisation[3] as well as the creation of building 'majors' mingling private development, engineering or utilities. The building sector remains strongly determined by national models (Campinos-Dubernet 1992). In most countries a few dominant firms have emerged, strengthening themselves through complex organisations and networks of subsidiaries. This is not far from the points made by Dominique Lorrain in his chapter (Lorrain 1995b) on the role of utilities or what may become of private development firms. Deregulation and the privatisation of urban utilities are slowly gaining ground all over Europe – a move that may offer more opportunities to private firms (Lorrain and Stoker 1995).

Thus, despite the fact that European countries are still consistently different from the USA, we have identified a whole range of private-sector actors who may play a more important role in the governance of urban areas should the opportunities arise, and if they are willing to participate. The point is that organised business interests have remained fairly weak in the majority of European cities, yet they should not be given our entire attention. All kinds of firms, networks, individuals, quangos and utilities exist in cities, in a complex organised environment.

The impact of these upon the governance of cities may vary enormously. Apart from organised interests themselves, these actors may be more or less organised, and their resources, autonomy, independent strategy and interests may differ. Secondly, they may be more or less localised or attached to the city. Their legitimacy to do anything will also vary immensely. A major suburban retailer may be very co-operative in order to enhance its local image and thus oppose shopkeepers' interests. Some multinationals may have a stake or an interest in a city project to improve the environment in which they operate. Moreover, the self-employed depend on the relative prosperity of the city. Firms may need certain resources that are peculiar to a city. Without being locked in a territory, they may take part in the production of collective consumer or capital goods. By contrast, firms or local associations may simply use local resources without any interest in the locality. Finally, scattered self-employed workers, interest groups and political rivalries may hinder any coalition-building. As a result, the combining of these characteristics may lead to different outcomes.

Private actors and urban governance in Europe

No country other than Britain has given so much importance to business interests in urban governance – a phenomenon which has prompted a significant amount of literature. However, the British case is the exception not the rule in Europe. In most countries, urban mayors have acquired more resources, more competence and more legitimacy (through direct election as has been the case recently in Italy and Germany). However, the logic of territorial competition is growing in importance in Europe, and pressure is often exerted in that direction through the mediation of private actors and organised business interests. However, the interaction between various actors, including local government and organised business groups or individual firms may assume different forms.

From government to fragmented local governance or more integrated urban governance?

Some authors argue for the British case that 'the incorporation of business leaders is associated with the disorganisation of local politics, a process leading to the consolidation of power in the hands of the central state' (Peck and Tickell 1995: 63). In other words, the trend towards the forced extension of market rationales supported by a strong centralist state leads towards a disorganised model of urban politics, a context which makes the political mobilisation of company leaders possible. The evidence from British case studies is compelling. However, the theoretical view behind these assumptions is both very stimulating and open to debate. It is also very UK-centred and could hardly be applied as such in other European countries.

A different theory may be put forward. First, the question of the disorganisation of urban politics under market restructuring (if we accept the view that national policies are usually more balanced towards cities than the brutal Conservative reorganisation in Britain) has to be taken seriously. In some cities, weak public and private actors and feeble organised interests tend to lead, under market pressure, to weak urban governance. In some major European cities, the strength of market regulation is such that there is extreme fragmentation in terms of urban governance such as in London, or to a lesser extent in Paris where the state plays an important role. But this fragmentation may not be so pronounced. These major urban regions are also among the more exposed to economic competition (Cheshire and Gordon 1995). In this context, certain employers' organisations, also strengthened by economic development, are able to articulate certain objectives and become involved, often in a loosely co-operative way, without serious processes of integration. Neoliberals argue that the process of governance should be shared between numerous highly motivated agencies, with an element of competition in order to increase efficiency, limit taxes, prevent heavy bureaucracies and to limit the role of social groups and politics. *The Economist* in London regularly sings the praises of London in comparison with other European cities, particularly Paris, claiming it complies with this model. In major cities where local government is fragmented and has little ability to represent the urban area, employers' organisations may play a more important role. In Paris, London or Rotterdam, the very strong and dynamic economic fabric is able to generate more resources and a stronger organisation. The Paris Chamber of Commerce and Industry has become a very rich, strong and influential organisation (it also runs France's most elite business schools). If competition between cities has an impact first and foremost on major international cities (or global cities as they are known (Sassen 1991)), then it is in the interest of private firms to promote major infrastructure development (transport, airports, hotels, exhibition centres) in order to be competitive and attract conferences or transnational firms' headquarters to new business districts (such as the London Docklands or La Défense). In the face of these changes, some social movements are likely to emerge to counteract the public policies which are inevitably related to this sort of fragmented governance with strong private actors. These are well described in Margit Mayer's chapter.

However, that is certainly not the whole picture of European cities. In contrast to what Peck and Tickell have argued, it may also be the case that in some cities market pressure leads to a reinforcement of the city as a collective actor (in certain circumstances that remain to be outlined). However, in that case, business interests and leaders are not mobilised as a result of the disorganisation of urban politics. On the contrary, they may mobilise in a context of *increased* urban governance, within the framework of processes of internal and external integration set up by various groups, firms, interests and institutions. Within the changing context described earlier on, economic interests are, in some cities, part of urban governance processes.

There again, such a sweeping statement requires some qualification. As Borraz (1999) rightly suggests, most of the research into interest groups, coalitions and governance tends to overemphasise the processes of integration within an urban area. However, in most cities, there is always a tension between the fragmentation in progress and the attempts being made to overcome it and to organise some sort of collective action. The point is not to take a naive view of the logic of integration, which is always difficult, partial and subject to change. However, as stated in the introduction, there is also a case for arguing that the micro view, which only considers interactions between actors, does not reveal the ways in which some cities have emerged as collective actors with relatively stable patterns of governance.

Interactions: how central are public–private partnerships to the creation of coalitions?

The whole question of interest formation and legitimacy in cities differs slightly from the way it is addressed at state or EU level. The strengthening of urban governance in most European countries (with strong variations between cities) has more to do with mobilisation, collective action and to negotiation than with domination and coercion. It follows that, in contrast to neo-corporatist literature, the whole issue of the legitimatisation of interest groups by public authorities is less central (but not absent) in cities. One is therefore unlikely to find examples of urban neocorporatism in the original sense of the term, or limited examples of the Schmitter/Streeck type of regulation through official associations. Interactions between economic interests and local authorities follow regular patterns both in terms of strategies and in terms of public policy.

In most cities, political leaders tend to have dialogue with private interest groups. But that is not saying much. In many cities, especially in Britain, we have seen the development of private–public partnerships in urban flagship projects, for example. This phenomenon has been studied in depth as it appeared in Germany, Spain and in France (see Heinz 1993). The 1980s' property boom proved to be a major driving force behind the development of large-scale urban projects. If Britain was at the forefront of this movement, in a context of scarce resources, state incentives and booming real estate, many cities developed some forms of partnership in the context of urban regeneration, for instance. These are, however, often one-off co-operative actions, structured around one key project. In cases where such partnerships flourish, one often finds weak public partners and social groups, as well as fragmented governance (British cities are a good illustration of this). This phenomenon is not so widespread in Europe though. Such partnerships and flagship projects were non-existent or very limited in Italy and most Spanish cities except Barcelona and Madrid. Most Scandinavian cities have resisted this trend. Only recently has the city of Helsinki accepted some form of public–private partnership to renovate an area of disused industrial land in the city centre, or to develop the harbour.

At the other end of the spectrum, some cities have managed to develop collective strategies which fully integrate business interest groups and their leaders. In some cases, Chambers or employers' organisations join forces to elaborate the various stages of a strategy for the city and are closely involved in its implementation. Often, and this is a new development, employers' organisations are mobilised around economic development issues especially in order to represent the city and its strategies *vis-à-vis* other firms, the state or other cities. In most cases, private actors are integrated into image-building operations to market the city to outside investors and the middle classes. The interaction between business groups and networks may be stabilised and encourage the development of sophisticated exchanges in terms of culture, property development, policy to combat social exclusion, football, transport, parking and land use. This does not necessarily prevent conflicts from occurring, but a local social system of action is created thanks to the development of regulation which can also be based on reciprocity and trust. Such cases are found, for instance, in some German cities such as Stuttgart, in some Italian cities such as Bologna or Milan, in some French cities such as Lyon and Rennes in the 1980s, Birmingham in the UK and Barcelona and Valencia in Spain. Strong governance systems aiming at the integration of employers' organisations may only have a feeble external impact, and give rise to conservative public policy which seeks to limit development, for instance in Strasburg or Bordeaux in the 1980s. In the UK, for instance, so-called anti-economic-growth coalitions, to use that phrase, have shown they have some impact in Swindon (Harloe 1992) or Norwich. In some European cities, local authorities had resisted the idea of entering into urban competition, and of organising collective strategies for economic development until very recently, in Copenhagen and Amsterdam (Harding 1996), Helsinki, Dublin and various Italian cities, for instance. Changes often occur when new political leaders are elected. The new generation of Italian mayors in Turin, Venice, Naples and Rome, for instance, is rapidly trying to develop collective strategies incorporating private-sector organisation.

Beyond the direct involvement of local business organisations, it may also be the case that strong urban governance leads some external actors such as multinational firms or major utilities firms to consider the city and its strategies in their development. Strong urban governance may orientate the strategies of various economic actors in a number of ways, including those suggested by Pierre Veltz in chapter 1. In France, for instance, as shown by Dominique Lorrain, Lyonnaise des Eaux or Générale des Eaux have long-term interests in some cities through the various services they provide (Lorrain 1995a). As argued by Lorrain, the global growth of a local industry brings large private firms to the fore in European cities.

This development may be analysed in the light of the US theories concerning urban regimes (Stone 1989; Elkin 1987) and urban growth coalitions (Logan and Molotch 1987), which particularly focus on: (1) the coalition-building process between the private sector and the local authority in a context of decentralised and market-oriented politics; and (2) on the structural financial dependence of cities upon the private sector. These concepts did not appear to hold their own entirely in a European context (Keating 1991; Stoker 1994; Harding 1995; Le Galès

1995). These concepts, which originated in America, very much hinge on the relationship between local public authorities and private interests, such as local entrepreneurs, private developers, people of independent means, bankers, landowners or business-sector elites. The structural dependence of American cities on firms (especially in fiscal terms) acts as a powerful mechanism for coalition-building. In the case of growth coalitions, the driving force lies mainly in property development issues. As for Britain, Harding (1991) and Keating (1991) among others have pointed out differences between Britain and the USA particularly as far as the private sector is concerned – differences that also apply to most Western European countries:

- There is only a very limited role for private firms in politics in Britain, compared with the USA, for instance in choosing candidates for local elections.
- British local authorities do not depend upon tax from private firms for their budget; they depend structurally upon central government fundings while American local authorities are very dependent upon firms.
- Land use regulations are also different, and in Britain public organisations, trusts, foundations and various levels of government own substantial amounts of land; regulations and planning regulations in particular are also different.
- With some exceptions, British financial institutions, major firms and organised interests are centralised and unlikely to be involved in local politics, even if things did change somewhat in the 1980s.
- Most importantly, as Keating argues (1991), in the case of Britain and France the state and the various bodies and organisations which are related to, or part of, the state form a complex set of networks and arrangements with local authorities. In other words, failing to look carefully at the relationships between local authorities and the state would be seriously misleading.

Britain, the most neoliberal country, is too often seen as the showcase of changing urban politics in Europe. One big change in the cast of players in urban politics in the UK, reflecting the growing importance of the politics of production, has been the increasing role played by private-sector utilities and transport groups (Walsh 1995; Graham and Marvin 1994). To some extent, this can be attributed to government policy change, and as there are now many opportunities for business leaders to take up leading positions in the new development agencies, public programmes have been re-orientated towards various forms of private funding and public–private partnerships. The privatisation programmes have created new urban 'players' in the form, for example, of companies who run formerly public utilities (gas, water, electricity, railways) and have a stock of urban assets they are strongly encouraged to develop. In the economic world, orientations were already changing irrespective of government prompting. A movement which gathered pace during the 1980s saw the major national employers' organisations encouraging their members to organise themselves better at urban level and contribute to local development initiatives and strategies. Many companies did so, for a variety of reasons, ranging from simple corporate philanthropy to self-interest geared to potential profit, political kudos, presenting a positive corporate image or redeeming the corporate conscience. Although many of the initiatives that emerged were very marginal and small in scale, they at least achieved a greater understanding by the business community of the pressures facing organisations in the public and voluntary sectors. With public-sector funding for non-statutory agencies declining rapidly, they also began to trigger direct private–voluntary-sector relationships which had previously been very unusual.

Private interests were analysed from this angle in the literature of urban politics. For instance, Stoker and Mossberger (1994) (and, to some extent, Keating 1991) have developed an urban regime framework, adapting it for comparative purposes. In the American urban regime theory, coalition-building mainly involves the elected members (and officers) and business representatives of local authorities. Even if some other groups are included in the process (such as community organisations or minority groups in Stone's account of Atlanta), on the whole they appear to be marginal. Mossberger and Stoker suggest that the whole notion of an urban regime should be developed according to the key principle of collective organisation and action. According to them two other sets of actors should be brought into a European context: (1) community interests, minorities, neighbourhoods, organised labour; and (2) professional officials employed by local government, local agencies or central and regional government. Even if cities have gained considerable autonomy and ability to obtain public resources and investments in most European countries, the fact remains that the way they organise and structure their relations with various parts of the state is important for their governance (despite the ongoing debate about how much this is still the case).

It would be opportune here to come back to an essential feature of urban regime theory, namely, the structural advantage enjoyed by business in capitalist societies. This structural advantage may vary considerably from one country to another, from one period to another and from one locality to another. In the growth coalition literature, it is often argued that a locality is more likely to resist expansionist pressure if it has enjoyed consistent growth and little unemployment (Harloe 1992). However, in most European countries, the state has imposed its own political and administrative regulations, and this has prevented local authorities from being directly involved with the market, thus protecting them from the rigours of market discipline. Until the mid 1970s, an urban regime approach would have proved irrelevant in most places. Centre–periphery relations were the order of the day. However, the relative retreat of the state has created openings for new opportunities.

Within the context of Europe, regardless of where cities stand regarding economic development, it is necessary to focus on their relative positions in terms of their relations with the state and their ability to obtain funds and public infrastructures or utilities. This brief analysis suggests that many combinations of coalitions are likely to emerge – perhaps more than in the USA – because the state is often fragmented and various state organisations have to be taken into account. In some countries, organisations associated with the tertiary sector also play an important role (as in Britain). Building a coalition that could find its stability in an urban regime is therefore a complex process, involving many actors. If collective action is the name of the game, it follows that one has to make use of all the analytical tools which enable one to go beyond the Olsonian paradox, including trust and reciprocity between actors, or identity and culture as resources for collective action. The creation of an urban regime in most European cities constitutes, as usual, a difficult coalition-building process, with competition between places as the only incentive in the absence of any structural dependence (at least partial) of local authorities' finances upon firms.[4] One way forward is to come back to the idea of examining the interaction between the state (central or local), markets and civil society, to examine how some cities or regions are structured around combinations of the market and social and political regulations which structure governance regimes.

Conclusion

Although there is little evidence of urban corporatism in European cities as such, and no clear pattern of reinforced organised urban business interests, private actors (individual and collective) are increasingly salient in the governance of cities. Globalisation processes and, to some extent, European integration, tend to reinforce the logic of competition between cities (Cheshire and Gordon 1996). In the most important urban areas, such as London, Paris and Berlin, that logic comes more readily into play. Other cities are also feeling the impact of economic restructuring, as urban poverty, for example, is on the increase nearly everywhere. It has been argued in this chapter that private actors are instrumental in encouraging the formation of competitive urban regimes. However, some private actors and interest groups may also feel threatened by globalisation and react accordingly. In the past, people of independent means and shopkeepers succeeded in preventing certain economic development projects in a number of cities. Industries were not accepted, for instance. In a similar way, local private developers and heads of medium-sized firms may feel fragile and try to reinforce the city as a collective actor to defend them and the society in which they live.

The organisation of business interests at city level is crucial if one is to understand the development of urban governance processes in European cities. Beyond the employers' associations and Chambers of Commerce, individual firms and various sorts of private actors may co-ordinate their strategies according to the prospects for their city in the short run or the long run. In some cities patterns of stable relations develop within a coherent strategy. However, seeing in a private–public partnership a substitute for local government and/or negotiated relations between social groups, community organisations, local councils and employers' organisations, as is sometimes suggested in some accounts of 'new urban governance', is grossly misleading and not even fully accurate in the British case.

However, the point here is to suggest that the structure of business interests in cities in European countries is to a certain extent related to or constrained by national traditions, and that it reveals consistent diversity which cannot be diminished by some uniform globalisation process. The actors are more or less local, organised and legitimised. They may have conflicts, greater or lesser access to coalitions and networks, or they can be excluded. In most European cities, the actors are often local authorities (including bureaucrats and politicians), state organisations, voluntary-sector organisations, various more or less public agencies (foundations, quangos, universities, hospitals), employers' and trade union organisations, organised social groups, major firms in urban utilities or private developers. Another dimension has to do with the formation of social groups. In other cities, the bourgeoisie may be structured as a social group in its own right both in terms of position within the labour market, property, way of life and reproduction (leisure, children's schools, consumption).

How do all these various factors co-exist in particular cities? There is no generic answer. One can only point to the crucial role played by cities as political actors in the construction of social groups and often in the leadership of local authorities, defining agenda, bringing alliances of public, private and voluntary-sector organisations together and overcoming internal conflicts. Beyond this observation it is difficult to point to any simple way in which interest group activity is guided and made coherent in European cities. As urban governance becomes more fragmented institutionally, external interest groups can benefit from more points of entry than ever before, and there is a much more complex system of interrelationships and dependencies between statutory agencies and non-statutory groups. That complexity and contingency are the order of the day is surprising only if one believes that simplicity and predictability are, or should be, the natural state of urban governance.

Notes

Some of the ideas in this chapter were developed jointly with Alan Harding. I also thank him for his comments.

1 Among major work: Berger (1981); Lehmbruch and Schmitter (1982); Golthorpe (1984), or recently Crouch (1993); Crouch and Streeck (1996).
2 Greenwood, Grote and Ronit (1992); Mazey and Richardson (1993); Mény, Muller and Quermonne (1995); Harvey (1995); Benington (1996).
3 See the various works of Elisabeth Campagnac: for instance, Campagnac (1992).
4 Although it is an interesting attempt, is it really useful to remain within the limits of the urban regime framework? The interesting thing is to look at the process of collective action and the various combinations of key actors including social groups.

References

Agulhon, M. (1988), *Histoire vagabonde*, vol. II: *Idéologies et politique dans la France du xixe siècle*, Paris: Gallimard.
Albert, M. (1991), *Capitalisme contre capitalisme*, Paris: Editions du Seuil.
Amin, A. (ed.) (1994), *Post-Fordism: A Reader*, Oxford: Basil Blackwell.
Andersen, S. and Munk, A. (1995), *The Welfare State Versus the Social Market Economy – Comparison and Evaluation of Housing Policies in Denmark and West Germany*, Scandinavian Housing and Planning Research.
Ascher, F. (1995), *Metapolis ou l'avenir des villes*, Paris: Odile Jacob.
Asmuth, G. (1996), 'Die Burgen denen, die drin wohnen', *Tageszeitung*, no. 3.
Atkinson, A., Rainwater, L. and Smeeding, T. (1995), *Income Distribution in OECD Countries*, Paris: OECD.
Augé, M. (1992), *Non-lieux*, Paris: Editions du Seuil.
Bagguley, P. (1986), 'Protest, Poverty and Power: A Case Study of the Anti-Poll Tax Movement', Leeds University.
Bagnasco, A. (1986), *Torino. Un profilo sociologico*, Turin: Einaudi.
 (1988), *La costruzione sociale del mercato*, Bologna: Il Mulino.
 (1988), *L'Italia in tempi di cambiamento politico*, Bologna: Il Mulino.
Bagnasco, A. and Negri, N. (1994), *Classi, ceti, personne. Esercizi di analisi sociale localizzata*, Naples: Liguori.
Bagnasco, A. and Sabel, C. (eds.) (1994), *PME et développement économique en Europe*, Paris: La Découverte (English translation: *Small Firms in Europe*, London: Pinter, 1995).
Bagnasco, A. and Trigilia, C. (1984), *Società e politica nelle aree di piccola impresa. Il caso di Bassano*, Venice: Arsenale.
 (1985), *Società e politica nelle aree di piccola impresa. Il caso della Val Delsa*, Milan: Franco Angeli.
La construction sociale du marché, Cachan: Presses de l'ENS.
Batten, D. F. (1995), 'Network cities: creative urban agglomerations for the 21st century', *Urban Studies*, 32: 313–27.
Baudant, A. (1980), *Pont-à-Mousson (1918–1939), stratégies industrielles d'une dynastie lorraine*, Paris: Publications de la Sorbonne.
Bauer, M. and Bertin-Mourot, B. (1995), 'La tyrannie du diplôme initial et la circulation des élites: la stabilité du modèle français', in Suleiman, E. and Mendras, H. (eds.) (1995), *Le recrutement des élites en Europe*, Paris: La Découverte.
Becattini, G. (ed.) (1987), *Mercato e forze locali: il distretto industriale*, Bologna: Il Mulino.

References

Benevolo, L. (1993), *La ville dans l'histoire européenne*, Paris: Editions du Seuil.

Benington, J. (1996), 'Partnership in social policy in the EU', Local Government Centre, Warwick: mimeo (forthcoming, 1999, London: UCC Press).

Bennett, R. J. and Krebs, G. (1991), *Local Economic Development: Public–Private Partnership Initiatives in Britain and Germany*, London: Belhaven.

Berger, S. (ed.) (1981), *Organizing Interests in Western Europe*, Cambridge: Cambridge University Press.

Berry, B. (ed.) (1976), *Urbanization and Counterurbanization*, Beverly Hills: Sage.

Bertels, L. and Nottenbohm, H.-G. (eds.) (1983), *Außer man tut es. Beiträge zu wirtschaftlichen und sozialen Alternativen*, Bochum: Germinal.

Bessy, P. (1990), *Typologie socio-professionnelle de l'Ile-de-France. 22 types de communes*, Paris: INSEE.

Bevilacqua, P. (1993), *Breve storia dell'Italia meridionale*, Rome: Donzelli Editore.

Beywl, W. (1983), 'Alternative Ökonomie-Modell zur Finanzierung von Selbsthilfeprojekten', in Bertels and Nottenbohm (1983).

— (1988), 'Stand und Perspektiven der Forschung zur Alternativökonomie', *Forschungsjournal Neue Soziale Bewegungen*, 2: 7–12.

BfLR, *Spatial Planning Policies in a European Context*, Bonn: Federal Ministry of Regional Planning.

Bidou, C. et al. (1983), *Les couches moyennes salariées. Mosaïque sociologique*, Paris: Rapport au ministère de l'urbanisme et du logement.

Bishop, M., Kay, J. and Mayer, C. (1994), *Privatization and Economic Performance*, Oxford: Oxford University Press.

Black, A. (1984), *Guilds and Civil Society in European Political Thought from the Twelfth Century to the Present*, London: Methuen.

Blattert, B., Rink, D. and Rucht, D. (1994), 'Von den Oppositionsgruppen der DDR zu den neuen sozialen Bewegungen in Ostdeutschland?', Science Center Berlin, FS III: 9–101.

Bodenschatz, H., Heiser, V. and Korfmacher, J. (1983), *Schluss mit der Zerstörung? Stadterneuerung und städtische Opposition in West-Berlin, Amsterdam und London*, Giessen: Anabas.

Body-Gendrot, S. (1992), *Ville et violence*, Paris: Presses Universitaires de France.

— (1995), 'Marginalization and political responses in the French context', paper presented at the ESF Conference.

Bonneville, M., Buisson, M. E. and Rousier, N. (1993), 'L'internationalisation des villes en Europe: un même défi, des processus différents', in Bonneville, M. (ed.) (1993), *L'avenir des villes. Excellence et/ou diversité*, Lyons: Programme Rhône-Alpes, pp. 85–106.

Borraz, O. (1998), *Le gouvernement des villes*, Rennes: PUR.

— 'Gouvernabilité des villes et action publique', in Balme, R., Faure, A. and Mabileau, A. (eds.) (1999), *Gouvernement local et action publique en Europe*, Paris: Presses de la FNSP.

Boyer, R. and Hollingsworth, R. (eds.), *Contemporary Capitalism*, Cambridge: Cambridge University Press.

Brotchie, J., Batty, M., Blakely, E., Hall, P. and Newton, P. (eds.) (1995), *Cities in Competition. Productive and Sustainable Cities for the 21st Century*, Melbourne: Longman.
Bruhns, H. (1995), 'M. Weber en France et en Allemagne', *Revue Européenne des Sciences Sociales*, 101: 107–21.
Brunet, J.-P. (1980), *Saint-Denis, la ville rouge. 1890–1939*, Paris: Hachette.
Brunet, R. (ed.) (1989), *Les villes européennes*, RECLUS-DATAR, Paris: La Documentation Française.
Burns, D. (1992), *Poll Tax Rebellion*, Stirling: AK Press.
Burtenshaw, D., Bateman, M. and Ashworth, G. J. (1991), *The European City, a Western Perspective*, London: David Fulton.
Caldeira, T. P. (1996), 'Un nouveau modèle de ségrégation spatiale: les murs de São Paulo', *Revue Internationale des Sciences Sociales*, 147: 65–77.
Campagnac, E. (ed.) (1992), *Les grands groupes de la construction: de nouveaux acteurs urbains?*, Paris: L'Harmattan.
Campbell, J., Hollingsworth, R. and Linberg, L. (eds.) (1991), *Governance of the American Economy*, Cambridge: Cambridge University Press.
Campinos-Dubernet, M. (1992), 'La diversité des bâtiments européens: l'incidence des modèles nationaux', in Campagnac (1992).
Caron, F. and Cardot, F. (eds.) (1991), *Histoire de l'électricité en France*, Paris: Fayard.
Castel, R. (1995), *Les métamorphoses de la question sociale. Une chronique du salariat*, Paris: Fayard.
Castells, M. (1973), *Luttes urbaines et pouvoir politique*, Paris: Maspero.
—— (1989), 'Social movements and the informational city', *Hitotsubaschi Journal of Social Studies*, 21: 197–206.
Cattacin, S. (1994), *Stadtentwicklungspolitik zwischen Demokratie und Komplexität. Zur politischen Organisation der Stadtentwicklung: Florenz, Wien und Zürich im Vergleich*, Frankfurt: Campus.
Cattan, N. (1993), 'La dynamique des échanges aériens internationaux entre les grandes villes européennes', *Revue d'Economie Régionale et Urbaine*, 4.
Cattan, N., Pumain, D., Rozenblat, C. and Saint-Julien, T. (1994), *Le système des villes européennes*, Paris: Anthropos.
Ceccarelli, P. (1982), 'Politics, parties, and urban movements: Western Europe', in Fainstein, N. and Fainstein, S. (eds.), *Urban Policy under Capitalism*, Beverly Hills: Sage.
Champion, A. (ed.) (1989), *Counterurbanization. The Changing Place and Nature of Population Deconcentration*, London: Arnold.
Chandler, A. D. (1977), *The Visible Hand. The Managerial Revolution in American Business*, Cambridge, MA: Harvard University Press, 1977.
Chenu, A. (1996), 'Les étrangers dans les agglomérations françaises', in Pumain, D. and Godard, F. (eds.) (1996), *Données urbaines*, Paris: Economica, pp. 225–34.
Chenu, A. and Tabard, N. (1993), 'Les transformations socioprofessionnelles du territoire français, 1982–1990', *Population*, 6: 1735–70.
Cherki, E. and Mehl, D. (1978), 'Urbane Bewegungen in der Region Paris', in Mayer, M. *et al.* (eds.) (1978), *Stadtkrise und soziale Bewegungen*, Frankfurt: EVA.

References

Cherki, E. and Wieviorka, M. (1978), '"Autoriduzione" in Turin', in Mayer, M. et al. (eds.) (1978), *Stadtkrise und soziale Bewegungen*, Frankfurt: EVA.

Cheshire, P. (1993), 'Explaining the recent performance of the European Community's major urban regions', *Urban Studies*, 27: 307–29.

—— (1995), 'A new phase of urban development in Western Europe? The evidence for the 1980s', *Urban Studies*, 32: 1045–63.

Cheshire, P. and Gordon, I. (eds.) (1995), *Territorial Competition in an Integrating Europe*, Aldershot, Hants: Avebury.

Cheshire, P., Hay, D., Carbonaro, G. and Bevan, N. (1989), *Urban Problems in Western Europe: An Economic Analysis*, London: Unwin Hyman.

Clark, W. A. V. (1991), 'Residential preferences and neighborhood racial segregation. A test of the Schelling segregation model', *Demography*, 28 (1).

Clarke, S. E. and Mayer, M. (1986), 'Responding to grassroots discontent: Germany and the United States', *International Journal of Urban and Regional Research*, 10: 401–17.

Coing, H. (1996), *Rénovation urbaine et changement social*, Paris: Editions Ouvrières.

Coleman, J. (1990), *Foundations of Social Theory*, Cambridge, MA: Belknap/Harvard University Press.

Coleman, W. and Jacek, H. (eds.) (1989), *Regionalism, Business Interests and Public Policy*, London: Sage.

Congdon, P. (1984), *Social Structure and Change in London Wards. Evidence from the 1971 and 1981 Censuses*, London: Greater London Council.

—— (1987), *A Map Profile of Change in London Wards*, London: London Research Centre.

de Coninck, F. (1995), *Travail intégré, société éclatée*, Paris: Presses Universitaires de France.

Conti, S. (193), *Effetto città*, vol. II: *L'Europa nella transizione post-socialista*, Turin: Fondazione G. Agnelli, 1993.

Conti, S., Dematteis, G. and Emanuel, C. (1995), 'The development of areal and network systems', in Dematteis, G. and Guarrasi, V. (eds.) (1995), *Urban Networks. Geo-Italy 2*, Bologna: Pátron, pp. 45–68.

Conti, S. and Spriano, G. (eds.) (1990), *Effetto città*, vol. I: *Sistemi urbani e innovazione: prospettive per l'Europa degli anni novanta*, Turin: Fondazione G. Agnelli, 1990.

Cooke, P. (ed.) (1988), *Localities*, London: Unwin and Hyman.

Coutard, O. (ed.) (1999), *The Governance of Large Technical Systems*, London: Routledge, forthcoming.

Crouch, C. (1993), *Industrial Relations and European State Traditions*, Oxford: Clarendon Press.

Crouch, C. and Streeck, W. (eds.) (1996), *Les capitalismes en Europe*, Paris: La Découverte.

—— (1997), (English translation: *Diversity of Capitalism*, London: Sage).

Crozier, M. and Friedberg, E. (1977), *L'acteur et le système*, Paris: Editions du Seuil.

Dackweiler, R., Poppenhusen, M., Grottian, P. and Roth, R. (1990), 'Struktur und Entwicklungsdynamik lokaler Bewegungsnetzwerke in der

Bundesrepublik: Eine empirische Untersuchung an drei Orten', Berlin: Research Report DFG, Freie Universität.
Dangshaft, J. and Ossenbrügge, J. (1990), 'Hamburg: crisis management, urban regeneration and social democrats', in Judd and Parkinson (1990), pp. 86–106.
Davezies, L. (1995), 'L'inégalité spatiale en France', in Savy, M. and Veltz, P. (eds.) (1995), *Economie globale et réinvention du local*, La Tour d'Aigues: DATAR/Editions de l'Aube.
(1996), 'Les produits des grandes villes françaises', in Pumain, D. and Godard, F., *Données urbaines*, Paris: Anthropos.
Davis, M., (1990), *City of Quartz*, London: Verso.
De Lavergne, F. and Mollet, P. (1991), 'The international development of intermediate sized cities in Europe: strategies and networks', *Ekistics*, 58: 368–81.
De Roo, P. (1994), 'Quatre scénarios pour les villes d'Europe entre réseau et territoire', in DATAR, *Dossier prospective et territoires*, Paris: La Documentation Française, pp. 77–96.
Deben, L. (1990), 'Urban landsquatting: another way of living in Amsterdam', Amsterdam University: Working Paper.
Dematteis, G. (1991), 'Possibilità e limiti dello sviluppo locale', Prato, IRIS, *Incontri Pratesi sullo sviluppo locale* (reprinted 1994 in *Sviluppo locale*, no. 1: 10–31).
(1996), 'Towards a metropolitan urban system in Europe: core centrality versus network distributed centrality', in Pumain, D. and Saint-Julien, T. (eds.) (1996), *Urban Networks in Europe*, Paris: John Libbey, pp. 19–28.
Desplanques, G. and Tabard, N. (1991), 'La localisation de la population étrangère', *Economie et Statistique*, 242.
Dunford, M. and Kafkalas, G. (eds.) (1992), *Cities and Regions in the New Europe*, London: Belhaven Press.
Dunleavy, P. (1991), *Democracy, Bureaucracy and Public Choice*, Hemel Hempstead: Harvester Wheatsheaf.
Duroy, S. (1996), *La distribution d'eau potable en France*, Paris: LGDJ.
Elkin, S. E. (1987), *City and Regime in the American Republic*, Chicago: University of Chicago.
Emmanuel, C. (1988), Le trasformazion: recenti delle reti urbane della Padania centro-occidentale, Ph.D. thesis, University of Pisa.
Esping-Andersen, G. (ed.) (1993), *Changing Classes. Stratification and Mobility in Post-Industrial Societies*, London: Sage.
(1996), *Welfare States in Transition*, London: Sage.
Fainstein, S., Gordon, I. and Harloe, M. (eds.) (1992), *Divided Cities*, Oxford: Blackwell.
Fainstein, S. and Hirst, C. (1994), 'Urban social movements', in Judge, D., Stoker, G. and Wolman, H. (eds.) (1994), *Theories of Urban Politics*, Newbury Park: Sage.
Fehse, W. (1995), *Selbsthilfe-Förderung – 'Mode' einer Zeit? Eine Prozeß- und Strukturanalyse von Programmen zur Unterstützung von Selbsthilfeaktivititäten*, Frankfurt: Peter Lang.

Fielding, A. J. (1982), 'Counterurbanisation in Western Europe', *Progress in Planning*, 17: 2–52.
Fourcaut, A. (1986), *Bobigny, banlieue rouge*, Paris: Editions Ouvrières-Presses de la FNSP.
Friedberg, E. (1983), *Le pouvoir et la règle (dynamiques de l'action organisée)*, Paris: Editions du Seuil.
Friedmann, J. (1986), 'The world city hypothesis', *Development and Change*, 17: 69–83.
Friedmann, J. and Wolff, G. (1982), 'World city formation: An agenda for research and action', *International Journal of Urban and Regional Research*, 6/3: 309–34.
Froessler, R. *et al.* (eds.) (1994), *Lokale Partnerschaften: die Erneuerung benachteiligter Quartiere in europäischen Städten*, Basel: Birkhäuser.
Fuà, G. and Zacchia, C. (eds.) (1983), *Industrializzazione senza frattura*, Bologna: Il Mulino.
Galgano, F. (1976). *Storia del diritto commericale*, Bologna: Il Mulina.
Garrard, J. (1992), 'Le scandale du gaz à Salford en 1887', *Flux*, 1: 9–24.
Georgin, C. (1931), *Cours de droit administratif*, Paris: Ecole Spéciale des Travaux Publics.
Girault, J. (1977), *Sur l'implantation du Parti communiste français dans l'entre-deux-guerres*, Paris: Editions Sociales.
Godard F. *et al.* (1973), *La rénovation urbaine à Paris. Structure urbaine et logique de classe*, Paris: Mouton.
Godts, X. (1978), 'Stadtsanierung und städtische Konflikte in Brüssel', in Mayer, M. *et al.* (eds.) (1978), *Stadtkrise und soziale Bewegungen*, Frankfurt: EVA.
Golthorpe, J. (ed.) (1984), *Order and Conflict in Contemporary Capitalism*, Oxford: Clarendon Press.
Gottmann, J. (1976), 'Megalopolitan systems around the world', *Ekistics*, 43: 109–13.
—— (1991), 'The dynamics of city networks in an expanding world, *Ekistics*, 58: 277–81.
Graham, S. and Marvin, S. (1994), 'Cherry picking and social dumping: British utilities in the 1990s', *Utilities Policy*, 4.
—— (1996), *Telecommunications and the City*, London: Routledge.
Greenwood, J., Grote, J. and Ronit, K. (eds.) (1992), *Organised Interests in the EC*, London: Sage.
Gregory, D. and Urry, J. (1985), *Social Relations and Spatial Structures*, London: Macmillan.
Hamnett, C. (1984), 'Housing the two nations. Sociotenurial polarization in England and Wales, 1961–81', *Urban Studies*, 43: 389–405.
—— (1987), 'A tale of two cities: sociotenurial polarisation in London and the South East, 1966–1981', *Environment and Planning A*, 19: 537–56.
—— (1995), 'Les changements socio-économiques à Londres', *Sociétés Contemporaines*, 22–23: 15–32.
Hannerz, U. (1980), *Explorer la ville*, Paris: Editions de Minuit
Harding, A. (1991), 'The rise of growth coalitions, UK style?', *Environment and Planning C: Government and Policy*, 9.

(1995), 'Elite theory and growth machines', in Judge, D., Stoker, G. and Wolman, H. (eds.) (1995), *Theories of Urban Politics*, London: Sage.

(1996) 'Urban regimes in five European cities', unpublished paper for the ESRC, mimeo.

Harding, A., Dawson, J., Evans, R. and Parkinson, M. (eds.) (1994), *European Cities towards 2000*, Manchester: Manchester University Press.

Harding, A. and Le Galès, P. (1997), 'Globalization and urban politics', in Scott (1997).

Harloe, M. (1992), 'Switching to the slow lane: restraining growth in a boom town', communication to the International Conference on comparative regional studies, Sendaé (Japan), 19–25 September.

(1995), *The People's Home? Social Rented Housing in Europe and America*, Oxford: Blackwell.

Harloe, M., Pickvance, C. and Urry, J. (eds.) (1998), *Do Localities Matter?*, London: Unwin Hyman.

Harvey, D. (1989), *The Conditions of Post-Modernity*, Oxford: Blackwell.

Harvey, J. (1995), *Networking in Europe*, London: NCVO.

Häußermann, H. and Siebel, W. (1993), 'Festivalisierung der Stadtpolitik. Stadtentwicklung durch große Projekte', special issue of *Leviathan*, 13.

Heinelt, H. and Mayer, M. (eds.) (1992), *Politik in Europäischen Städten*, Berlin: Birkhäuser.

Heinz, W. (ed.) (1993), *Partenariats public-privés dans l'aménagement urbain*, Paris: L'Harmattan.

Hervieu, B. (1993), *Les champs du futur*, Paris: F. Bourin.

Hirshman, A. (1981), *The Passions and the Interests: Political Argument for Capitalism before its Triumph*, Princeton, NJ: Princeton University Press.

Hoffmann-Martinot, V. (1995), 'La relance du gouvernement métropolitain en Europe, le prototype de Stuttgart', *Revue Française d'Administration Publique*.

Hoggett, P. and Burns, B. (1991–2), 'The revenge of the poor: The anti-poll tax campaign in Britain', *Critical Social Policy*, 10.

Hohenberg, P. M. and Lees, L. H. (1985), *The Making of Urban Europe*, Cambridge, MA: Harvard University Press.

Hughes, T. (1983), *Networks of Power: Electrification in Western Society, 1880–1930*, Baltimore: Johns Hopkins University Press.

(1996), 'Les immigrés jugent la France', *Le Nouvel Observateur*, 1667, 17–23: October.

Jacobs, B. D. (1992), *Fractured Cities. Capitalism, Community and Empowerment in Britain and America*, London: Routledge.

Jacquier, C. (1991), *Voyage dans dix quartiers en crise*, Paris: L'Harmattan.

Jayet, H., Puig, J. P. and Thisse, J. F. (1996), 'Enjeux économiques de l'organisation du territoire', *Revue d'Economie Politique*, 106: 128–56.

Jessop, B. (1994), 'Post-Fordism and the state', in Amin (1994), pp. 251–79.

Judd, D. and Parkinson, R. (eds.) (1990), *Leadership and Urban Regeneration*, London: Sage.

Julien P. (1994), 'La métropolisation des emplois', *INSEE-Première*, 349.

Julkunen, R. (1992), *Hyvinvointivaltio käännekohdassa* ('Welfare State at the Turning Point'), Tampere: Vastapaino.

Kaijser, A. (1995), *Striking Bonanza: The Establishment of a Natural Gas Regime*

References

in the Netherlands, Conférence grands systèmes et réseaux techniques, Noisy-le-Grand: LATTS.

Kasvio, A. (1994), *Uusi työn yhteiskunta* ('New Work Society'), Jyväskylä: Gaudeamus.

Kautto, M., Heikkilä, M., Hvinden, B., Marklund, S. and Ploug, N. (1999), *Nordic Social Policy: Changing Welfare States*, London: Routledge.

Keating, M. (1991), *Comparative Urban Politics. Power and the City in the United States, Canada, Britain, and France*, Aldershot, Hants: Edward Elgar.

King, A. D. (1990), *Global Cities*, London: Routledge.

King, D. (1987), *The New Right: Politics, Markets and Citizenship*, London: Macmillan.

Knorr-Siedow, T. and Willmer, W. (1994), *Sozialverträglicher Umgang mit unkonventionellen, mobilen Wohnformen am Beispiel des Wohnens in Wohnwagendörfern oder Wagenburgen*, Berlin: IRS.

Knox, P. and Taylor, P. (eds.) (1995), *World Cities in a World System*, Cambridge: Cambridge University Press.

Konrad, G. (1996), 'Konrad, l'esprit de dissidence', interview with M. Van Renterghem, *Le Monde*, 7 November.

Krämer-Badoni, T. (1990), 'Die Dethematisierung des Sozialen – Ansätze zur Analyse städtischer sozialer Bewegungen', *Forschungsjournal Neue Soziale Bewegungen*, 4: 20–7.

Krämer-Badoni, T. and Söffler, D. (1994), 'Die Rolle der städtischen Bürgerinitiativen in Westdeutschland und Ostdeutschland bei der Ausprägung lokaler Demokratie. Universität Bremen', working paper, ZWE Arbeit und Region, no. 13.

Kriesi, H. (1984), *Die Zürcher Bewegung*, Frankfurt: Campus.

Kunzmann, K. and Wegener, M. (1991), 'The pattern of urbanization in Western Europe', *Ekistics*, 58: 282–91.

Lang, S. (1995), 'Civil society as gendered space. Institutionalization and institution building within the German women's movement', in Scott, J. W. and Caplan, C. (eds.) (1995), *Translations, Environments, Transitions: The Meanings of Feminism in Changing Political Contexts*, London: Routledge.

Lankinen, M. (1994), *Taantuvatko lähiöt? Pääkaupunkiseudun lähiöt sosiaalisen segregaation valossa*, Ympäristöministeriö, Yhdyskuntasuunnittelun ja rakennetutkimuksen neuvottelukunnan julkaisu 3/9, Helsinki.

Lash, S. and Urry, J. (1994), *The Economies of Signs and Space*, London: Routledge.

Lauterbach J. (1994), 'Staats- und Politikverständnis autonomer Gruppen in der BRD', *Widersprüche*, 50: 101–19.

Le Bras, H. (1996), *Le peuplement de l'Europe*, Paris: La Documentation Française.

Le Galès, P. (1993), *Politique urbaine et développement local. Une comparaison franco-britannique*, Paris: L'Harmattan.

(1995), 'Du gouvernement local à la gouvernance urbaine', *Revue Française de Sciences Politiques*, 45.

(1998), 'Regulations and governance in European cities', *International Journal of Urban and Regional Research*, 22: 482–506.

Le Galès, P. and Harding, A. (1996), 'Villes et états en Europe', in Wright and Cassese (1996).
(1998), 'Cities and state', *West European Politics*, 1.
Le Galès, P. and Lequesne, C. (eds.) (1997), *Regions in Europe*, London: Routledge.
Le Galès, P. and Oberti, M. (1995), 'Les classes moyennes en Italie, Grande-Bretagne, France', Paris: MRT Report.
Leal Maldonado, J. (1990a), *La segregacion social en Madrid*, Madrid: Ayuntamiento de Madrid.
(1990b) 'Crecimiento economico y desigualdad social en la Comunidad de Madrid', *Econom'a y Sociedad*, 4: 55–66.
Lefevre, Ch. (1992) 'Collectivités locales et gouvernement des aires métropolitaines', Capacitation Thesis, Paris: Val de Marne University, Créteil.
Lehmbruch, G. and Schmitter, P. (eds.) (1982), *Patterns of Corporatist Policy-Making*, London: Sage.
Lehto, J. (1997), *Paikkalliser sosiaalipolitiikan tarve ja edellytykset*, Helsinki: Stakes.
Leibfried, S. and Pierson, P. (eds.) (1995), *European Social Policy. Between Fragmentation and Integration*, Washington, DC: Brookings Institution.
Lo, C. Y. H. (1990), *Small Property versus Big Government. Social Origins of the Property Tax Revolt*, Berkeley: University of California Press.
Logan, J. R., Alba, R. D. and McNulty, T. L. (1995), 'Les minorités des villes globales: New York et Los Angeles', *Sociétés Contemporaines*, 22–23: 69–88.
Logan, J. R. and Molotch, H. L. (1987), *Urban Fortunes. The Political Economy of Place*, Berkeley: University of California Press.
Lopez, R. (1996), 'Hautes murailles pour villes de riches', *Le Monde Diplomatique*, no. 50, March.
Lorrain, D. (1989), 'La montée en puissance des villes', *Economie et Humanisme*, 305: 6–21.
(1990), 'Le modèle français de services urbains', *Economie et Humanisme*, 312: 39–58.
(1993), 'Les services urbains, le marché et le politique', in Martinand, C. (ed.), *Le financement privé des équipements publics*, Paris: Economica, pp. 3–33.
(1995), 'L'extension du marché', in Lorrain and Stoker.
(1995b), (ed.), *Gestions urbaines de l'eau*, Paris: Economica.
Lorrain, D. and Stoker, G. (eds.) (1995), *Les privatisations urbaines en Europe*, Paris: La Découverte. (English translation *Urban Privatisation in Europe*, London: Pinter, 1996).
Lutter, H. (1994), 'Accessibility and regional development of European regions and the role of transport systems', *Spatial Development in Europe Research Colloquium*, 17–18 March, Ministry of the Environment, Copenhagen.
Maloutas, T. (1993), 'Social segregation in Athens', *Antipode*, 25: 223–39.
(1995), 'Ségrégation sociale et relations familiales dans deux villes grecques: Athènes et Volos', *Sociétés Contemporaines*, 22–23: 89–106.
Martin, R. (1984), *Patron de droit divin*, Paris: Gallimard.
Martinotti, G. (1993), *Metropoli: La nuova morfologia della città*, Bologna: Il Mulino.

Maurice, M. (1994), 'Acteurs, règles et contextes', *Revue Française de Sociologie*, 35: 645–58.

Mayer, M. (1993), 'The role of urban social movement organisations in innovative urban policies and institutions', in Getimis, P. and Kafkalas, G. (eds.) (1993), *Urban and Regional Development in the New Europe*, Athena: Topos.

—— (1994), 'Post-Fordist city politics', in Amin (1994).

—— (1995) 'Urban governance in the post-Fordist city', in Healey, P. *et al.* (eds.) (1995), *Managing Cities, the New Urban Context*, Chichester: John Wiley.

—— (1996), 'Neue Kooperations- und Handlungsansätze in der Stadtentwicklungspolitik im deutsch-amerikanischen Vergleich', Research Project, FNK Berlin.

Mayer, M., Jahn, W. and Sambale, J. (1995), *Stadtentwicklung und Obdachlosigkeit in Berlin zwischen globalen Zwängen und lokalen Handlungsoptionen*, Research Programme on Berlin, Berlin Free University.

Mazey, S. and Richardson, J. (eds.) (1993), *Lobbying in the European Community*, Oxford: Oxford University Press.

McArthur, A. (1995), 'The active involvement of local residents in strategic community partnerships', *Policy and Politics*, 23: 61–71.

Meijer, M. (1993), 'Growth and decline of European cities: the changing position of cities in Europe', *Urban Studies*, 30: 981–90.

Mendras, H. (1988), *La seconde révolution française*, Paris: Gallimard (2nd edition).

Mendras, H. and Schnapper, D. (eds.) (1990), *Six manières d'être européens*, Paris: Gallimard.

Mény, Y., Muller, P. and Quermonne, J. L. (eds.) (1995), *Public Policies*, London: Routledge.

Mingione, E. (ed.) (1993), 'The new urban poverty and the underclass', special issue of *International Journal of Urban and Regional Research*, 17.

Mollenkopf, J. and Castells, M. (eds.) (1989), *Dual City: Restructuring New York*, New York: Russell Sage.

Moriconi-Ebrard, F. (1993), *L'urbanisation du monde*, Paris: Anthropos.

Nanetti, R. (1988), *Growth and Territorial Politics: The Italian Model of Social Capitalism*, London: Pinter.

Narr, W.-D. *et al.* (1981), 'Berlin, Zürich, Amsterdam – Politik, Protest und die Polizei', *CILIP (Civil Liberties and Police)*, 9/10: 2–156.

Newman, P. and Thornley, A. (1996), *Urban Planning in Europe*, London: Routledge.

Nord (1994), *Yearbook of Nordic Statistics 199*, Århus.

Oberti, M. (1995), 'L'analyse localisée de la ségrégation urbaine', *Sociétés Contemporaines*, 22–23, 127–44.

—— (1996), 'La relégation urbaine: regards européens', in Paugam, S. (ed.) (1996), *L'exclusion: l'état des savoirs*, Paris: La Découverte, pp. 237–247.

O'Brien, R. (1992), *Global Financial Integration: The End of Geography*, London: Royal Institute of International Affairs.

OCS, (1982), *Cahiers d'observation du changement social*, Paris: CNRS, 18 vols.

OCS, (1987), *L'esprit des lieux*, Paris: CNRS.

Offe, C. (1985), 'New social movements: challenging the boundaries of institutional politics', *Social Research*, 52: 817–68.

Offe, C. and Wiesenthal, H. (1980), 'Two logics of collective action. Theoretical notes on social class and organisational form', *Political Power and Social Theory*, 1.
Offner, J. M. and Pumain, D. (eds.) (1996), *Réseaux et territoires, significations croisées*, La Tour d'Aigues: Editions de l'Aube.
Öresjö, E. (1995), 'Den Sönderfallande Staden' ('The Disintegrating City'), *Socialvetenskapliga tidskrift*, 2: 150–60.
Park, R. E. (1925), 'The city', in Park, R. E. and Burgess, E. W. (eds.) (1984), *The City: Suggestions for Investigation of Human Behaviour in the Urban Environment*, Chicago: Chicago University Press.
Parkinson, M., Bianchini, F., Dawson, J., Evans, R. and Harding, A. (1992), *Urbanisation and the Functions of Cities in the European Community*, Brussels: European Commission (DG XVI).
Peck, J. and Tickell, A. (1995), 'Business goes local: dissecting the business agenda in Manchester', *International Journal of Urban and Regional Research*, 19.
Perrier, J.-L. (1997), 'La vie artistique de Budapest perturbée par la loi du marché', *Le Monde*, 26 February.
Pinçon, M. and Pinçon-Charlot, M. (1989), *Dans les beaux quartiers*, Paris: Editions du Seuil.
(1992), *Quartiers bourgeois, quartiers d'affaires*, Paris: Payot.
Pinçon-Charlot, M., Preteceille, E. and Rendu, P. (1986), *Ségrégation urbaine. Classes sociales et équipements collectifs en région parisienne*, Paris: Anthropos.
Pontusson, J. (1996), 'Le modèle suédois en mutation, vers le néolibéralisme ou le modèle allemand?', in Crouch and Streeck (1996).
Portes, A. and Bach, R. (1985), *The Latin Journey. Cuban and Mexican Immigrants in the United States*, Berkeley: University of California Press.
Preteceille, E. (1993), *Mutations urbaines et politiques locales*, vol. II, Paris: CSU.
(1995), 'Division sociale de l'espace et globalisation. Le cas de la métropole parisienne', *Sociétés Contemporaines*, 22–23: 33–67.
Preteceille, E. and Pickvance, C. (eds.) (1991), *State Restructuring and Local Power*, London: Frances Pinter.
Preterceille, E. and Ribeiro, Luiz Cesar de Q. (1999), 'Tendências da Segregação social em metrópoles globais e designais: Paris e Rio de Janeiro nos anos 80', *Rivista Brasileira de Ciencias Sociais*, 14/40, 143–62.
Pumain, D. and Rozenblat, C. (1993), 'The location of the multinational firms in the European urban system', *Urban Studies*, 30.
Pumain, D., Rozenblat, C. and Moriconi-Ebrard, F. (1996), 'La trame des villes en France et en Europe', in Pumain, D. and Godard, F. (1996), *Données urbaines*, Paris: Anthropos.
Putnam, R. (1993), *Making Democracy Work*, Princeton, NJ: Princeton University Press.
Raffoul, M. (1996), 'Brésil sous cloche à São Paulo', *Le Monde Diplomatique*, 504, March 1996.
Rémy, J. (ed.) (1995), *Georg Simmel, ville et modernité*, Paris: L'Harmattan.
Rhein, C. (1994), 'La division sociale de l'espace parisien et son évolution', in Brun, J. and Rhein, C. (eds.) (1994), *La ségrégation dans la ville*, Paris: L'Harmattan, pp. 229–57.

References

Rhodes, M. (ed.) (1995), *Regions in the new Europe*, Manchester: Manchester University Press.

Ritter, K. (1822), *Die Erdkunde*, vol. I, Berlin: Reiner (2nd edition).

Rodwin, L. and Sazanami, H. (eds.) (1991), *Industrial Changeand Regional Economic Transformation*, London: HarperCollins.

Rondin, J. (1985), *Le sacre des notables. La France en décentralisation*, Paris: Fayard.

Roth, R. (1994), 'Lokale Bewegungsnetzwerke und die Institutionalisierung von Neuen Sozialen Bewegungen', in Neidhardt, F. (ed.) (1994), *Öffentlichkeit, öffentliche Meinung, soziale Bewegungen*, Opladen: Westdeutscher Verlag, pp. 13–36.

Rozenblat, C. (1996), 'La mise en réseau des villes au niveau européen', in Pumain, D. and Saint-Julien, T. (eds.) (1996), *Urban Networks in Europe*, Paris: John Libbey, pp. 85–101.

Salais, R. and Storper, M. (1993), *Les mondes de production. Enquête sur l'identité économique de la France*, Paris: Editions de l'EHESS.

Sallez A. and Verot P. (1991), 'Strategies for Cities to Face Competition in the Framework of European Integration', *Ekistics*, 58.

Salsbury S. (1995), 'Grands réseaux techniques, modèles de développement dans le temps', *Flux*, 22: 31–42.

Sassen, S. (1991), *The Global City: New York, London, Tokyo*, Princeton, NJ: Princeton University Press.

Saunders, P. (1986), *Social Theory and the Urban Question*, London: Hutchinson (2nd edition).

Schelling, T. (1978), *Micro-Motives and Macro-Behavior*, New York: W. W. Norton.

Schubert, D. (1990), 'Gretchenfrage Hafenstraβe. Wohngruppenprojekte in Hamburg', *Forschungsjournal Neue Soziale Bewegungen*, 3/4: 35–43.

Scott, A. (ed.) (1997), *The Limits to Globalization*, London: Routledge.

Scudo, G. (1978), 'Die 'Squatter' – Bewegung in Grossbritannien', in Mayer, M. et al. (eds.) (1978), *Stadtkrise und soziale Bewegungen*, Frankfurt: EVA.

Selle, K. (1991), *Mit den Bewohnern die Stadt erneuern. Der Beitrag intermediärer Organisationen zur Entwicklung städtischer Quartiere. Beobachtungen aus sechs Ländern*, Dortmund/Darmstadt: Dortmunder Vertrieb für Bau- und Planungsliteratur.

—— (1994), *Expositionen. Eine Weltausstellung als Mittel der Stadtentwicklung?*, Dortmund/Hanover.

Sennett, R. (1992), *La ville à vue d'œil*, Paris: Plon.

Simmel, G. (1965), 'Les grandes villes et la vie de l'Esprit', in Choay, F., *L'urbanisme, utopies et réalités*, Paris: Editions du Seuil (original edition 1903).

Simon, P. (1997), 'La statistique des origines: l'ethnicité et la 'race' dans les recensements aux Etats-Unis, Canada, et Grande-Bretagne', *Sociétés Contemporaines*, 21: 11–44.

Soja, E. and Scott, A. (eds.) (1996). *The City*, Berkeley: University of California Press.

Soldatos, P. (1991), 'Strategic cities, alliances: a comparison of North America with the European Community', *Ekistics*, 58: 346–50.

Sonobe, M. and Machimura, T. (1996), 'Globalization effect or bubble effect?

Social polarization in Tokyo', Paper for the Global Cities Project Seminar, New York: CUNY, October.
Stoker, G. (1994), 'Regime theory and urban politics', in Judge, D., Stoker, G. and Wolman, H. (eds.) (1994), *Theories of Urban Politics*, Newbury Park: Sage.
Stoker, G. and Mossberger, K. (1994), 'Urban regimes in comparative perspective', *Environment and Planning C: Government and Policy*, 12.
Stone, C. N. (1989), *Regime Politics. Governing Atlanta, 1946–1988*, Lawrence, KS: University of Kansas Press.
Stracke, E. (1980), *Stadtzerstörung und Stadtteilkampf in Frankfurt am Main*, Cologne: Pahl-Rugenstein Verlag.
Streeck, W. (1992), *Social Institutions and Economic Performances*, London: Sage.
Streeck, W. and Schmitter, P. (eds.) (1985), *Private Interests Government: Beyond Market and State*, Beverly Hills: Sage.
Swedberg, R. (1987), *Current Sociology*, London: Sage.
Swyngedouw, E. A. (1992), 'The Mammon quest. "Glocalisation", interspatial competition and the monetary order: the construction of new scales', in Dunford and Kafkalas (1992).
Tabard, N. (1993), 'Des quartiers pauvres aux banlieues aisées: une représentation sociale du territoire', *Economie et Statistique*, 270: 5–22.
Tarr, J. (1995), 'Transforming an energy system. The evolution of the manufactured gas industry and the transition to natural gas', unpublished paper to the Large Systems and Utility Networks Conference, Noisy-le-Grand: LATTS.
Therborn, G. (1995), *European Modernity and Beyond*, London: Sage.
Tilly, C. (1992), *Contraintes et capital dans l'Europe 990–1990*, Paris: Aubier.
Tilly, C. and Blockman, W. (eds.) (1994), *Cities and the Rise of the State in Europe*, Boulder, CO: Westview Press.
Todd, E. (1994), *Le destin des immigrés. Assimilation et ségrégation dans les démocraties occidentales*, Paris: Editions du Seuil.
Toubon, J.-C. and Messama, K. (1988), *La Goutte d'Or: constitution, modes d'appropriation et de fonctionnement d'un espace pluri-ethnique*, Paris: IAURIF.
Trigilia, C. (1985), 'La regolazione localistica: economia e politica nelle aree di piccola impresa', *Stato e Mercato*, 14.
—— (1991), 'A paradox of a region: economic regulation and the representation of interests', *Economy and Society*, 20: 306–27.
—— (1996), *Sviluppo senza autonomia*, Bologna: Il Mulino.
United Nations Centre for Human Settlements (1996), *An Urbanizing World. Global Report on Human Settlements 1996*, Oxford: Oxford University Press.
Van den Berg, L., Drewett, R., Klaassen, L. H. et al. (1982), *Urban Europe: A Study of Growth and Decline*, Oxford: Pergamon Press.
Van Waarden, F. (1989), 'Territorial differentiation of markets, states and business interest associations: a comparison of regional business associability in nine countres and seven economic sectors', in Coleman and Jacek (1989).
Veltz, P. (1983), *Mondialisation, villes et territoires. L'économie d'archipel*, Paris: Presses Universitaires de France.
Vergès, P. (1983), 'Approches des classes sociales', *Sociologie du Travail*, 2.

References

Wagner, D. (1993), *Checkerboard Square. Culture and Resistance in a Homeless Community*, Boulder, CO: Westview Press.

Walsh, K. (1995), 'Les marchés et le service public', in Lorrain and Stoker (1995).

Waters, S. (1995), 'The Chambers of Commerce and local development in France and Italy', unpublished Ph.D. dissertation, University of Limerick.

Weber, E. (1983), *La fin des terroirs, la modernisation de la France rurale*, Paris: Fayard.

Weber, M. (1921), *La ville*, Paris: Aubier-Montaigne, 1982 (original edition 1921).

Wilson, K. and Portes, A. (1980), 'Immigrant enclaves: an analysis of the labor market experiences of Cubans in Miami', *American Journal of Sociology*, 88: 295–319.

Wilson, W. J. (1980), *The Declining Significance of Race*, Chicago: University of Chicago Press.

—— (1987), *The Truly Disadvantaged. The Inner City, the Underclass and Public Policy*, Chicago: University of Chicago Press.

Wright, V. (ed.) (1993), *Les privatisations en Europe*, Arles: Actes Sud.

Wright, V. and Cassese, S. (eds.) (1996), *La recomposition de l'Etat en Europe*, Paris: La Découverte.

Zukunft Bauen, (1994), *Zukunft Bauen 1993–1994*, Berlin: Geisel Druck.

Index

Agnelli family 186
Agulhon, Maurice 171
air traffic, growth of 37; as measure of flows in urban network 65
Allarde Decree 162
Amiens, social divisions in 81
Amsterdam 1, 60, 68, 192; and elected mayors 188; growth in air traffic 37; housing policy, alternative 152 n.7; squatter movement 137
archipelago economy 33
Argentina, progress to service society 113
Athens 58; housing market 87; social segregation in 79
Austria, and business organisation 182
'Autonomous' [German protest movement] 143
auto-poiesis, theory of 64

Baltimore 157
banks, structure of 187
Barcelona 185, 191, 192; as European metropolis 59; urban policy 28
Baudant, Alain 165, 166–7, 168, 169
Belgium, progress to service society 113
Benevolo, L., 8–9
Bergamo 109
Berlin 3, 5, 11, 196; as European metropolis 58; housing policy 134, 137–8; social protest movements, 'Wagenburgen' 144–5; squatter movement 136–8, 152 n.8, 152 n11
Berry, B. 52
Bilbao 60
Birmingham 5, 105, 185, 192
Bologna 192; and urban policy 28
Bordeaux 192; foreign population 83
Bourdieu, Pierre 77
Bradford 105
Braudel, Fernand 34, 38
Britain: anti-poll tax movement 140; as class society 105; economy, characteristics of 18; gas supply, history of 171–4; mobility 36; Radcliff–Maud Report 153; social structure of medium-sized cities 104–6; urban social movements 152 n.6
Britain; business organisation 182–3, 184; involvement in urban governance 186, 188, 189; political influence of 184; political involvement 194; compared with USA 193
Bruges 5
Brunet, R. 58
Brussels 68; as European metropolis 59
Bucher, K. 34
Budapest, as European metropolis 59; political continuity in 96
building industry, and local governance 187–8
business, and urban governance 180–97

Cairo 2
Caisse de crédit aux départements et aux communes 168
California 2
Canterbury 105
Cantillon, Richard 33
capitalism, and historical development of cities 5; and nation-state 19; and cities 20; developments in modern 16–21; development of superior stage of 39
Cardiff 105
Castells, M. 11
Cattan, N. 10
central government, influence of 13
centralisation, urban 53–4
Chambers of Commerce 182–5
Charleroi 60
Chicago school 3, 75, 76, 85
Chile, progress to service society 113

Index

Christaller, W. 50
circles of production 45–6
cities: Americanisation of 78; as ameliorators of market forces 35; changes in organisation and functions 6; changes in structure of 64–5; characteristics of European 8–9, 11–12, 35; classification of 58, 59, 60–2; competition between 179, 195–6; concentration of 50, 68; conditions for development of 65; contemporary significance of 6, 7, 50; continued importance of European 15–16; definition 5, 58; density of 53; dynamism of smaller 92; hierarchical structure 50; history 5–7, 8–11, 74, 180–11; and identity 25, 29–30, 62, 92, 98; international functions 60; life cycle model 53–4; longevity of European 9–11; loss of significance 26; nodes in a network 49, 62, 63, 65; political activity in 92–4; political importance 28–9, 74; productivity 36; restructuring of 59–60, 72; stability of 49; strategic functions of 59; uniformity 64; unit of political and social organisation 25; weakening of traditional bonds 62–3. *See also* class, housing, local government, metropolises, politics, segregation, social division, social structure, urban governance
cities, types of 99; industrial cities 99–100; cities of producers 101–102; service-based cities 102–3; public-service-dependent 103–4
'civicness' 24
class: division, in cities, 76–81; and segregation 14, 88; structure, development of 223; and city development 14; and social structure 100; and tertiary proletariat 79; as tool for analysis of social segregation 76–7; problems of definition 76. *See also* middle class
Clermont-Ferrand, social divisions in 81
Cologne 60
Comité hygiène et eau 168
commuting, growth in 54
Compagnie Générale des Eaux 170
company organisation, changes in 54–5
competition, changed nature of economic 43; and locality 43–4; and skilled labour 423; economic 39; economic effects of 39–40
complexity, theories of 64

conservation, and cities 15
consumer choice, and segregation 88–91
conurbations, characteristics of European 11
Copenhagen 60, 192; as European metropolis 58; unemployment 125
counter-urbanisation 52–3, 54; and post-industrial phase 54
Coventry 105
culture, and nation state 17; regional 17

Davezies, Lucien 36
decentralisation 10; of welfare state 127–8; urban 53–4; decentralised concentration 55
Denmark: diminishing role for local authorities 128; progress to service society 113; unemployment 118, 120
diversification 64
Dreux 107
Dubai 33
Dublin 59, 60, 192
Duisberg 60
Duroy, Stéphane 170

economic development; and nation state 16–17; localised 33; role of banks 187
economic globalisation, *see* globalisation
economic growth, concentration of 35–6
economic organisation, changes in 40
economic polarisation 33
economic regulation 16–17, 29; types of 17–18
economic specialisation, reduction in 36
economic stability, conditions for 18–19
economies, Anglo-American and European models 17–18
economies, convergence of 36
economy, changes in, and the city 33–46; characteristics of continental 18
Edinburgh 60
education, and self-segregation 90
elections, abstention rates in municipal 92–3
employers' organisations 182–5
employment: and welfare state 115; changes in nature of 40; concentration of 35–6; decline in cities 54; growth in small towns 54; patterns in European cities 13, in Scandinavia 117–18; peak of industrial 113; publicly funded 37
Engels, F. 75

environmental protest movements 139–40
Essen 60
Esso 163, 164
European Union 8; and urban network 69; and urban planning 70; as response to globalisation 127

Fainstein, S. 11
Fielding, A. 53
Finland: diminishing role for local authorities 128; progress to service society 113; unemployment 118, 120; urbanisation of 117
firms, and impact of globalisation 39–40; changes in organisation 54–5
flexibility, search for economic 44
Florence 1, 181, 184
flows, measurement of 65–6
France: business organisation 182-2; composition of business leadership 187; distribution of foreigners 83; economic concentration in cities 36–7; effects of globalisation 39; industrial patterns 41; mobility 36; municipal election abstention rates 92–3; political activity in cities 93; provision of water supply 162, 165–71; public employment 37; racial enclaves 90; regional economic policy 36; social divisions within 81; social structure of medium-sized cities 106–7; urban social movements 133
Frankfurt 1, 60, 68; as European metropolis 59; growth in air traffic 37; squatter movement 133–4, 137, 151 n.1
Friedberg, E. 154

Garrard, J. 171, 173
gas, history of urban supply 157–61; British example 171–4; coal gas 157–8; coke gas 159; natural gas 159–60; Netherlands, development of urban gas supply 163–5; transformation of industry 160–1; water gas 158–9
General Association of Hygienists and Municipal Technicians 167. *See also* Propex
Générale des Eaux 192
Geneva 60
Genoa 5
Germany: business organisation 182; effects of globalisation 39; elected mayors 28; housing policy 136–8; progress to service society 113; urban policy 28; urban social movements 133–6, 136–8, 143, 144–5
ghettos, rarity in Europe 13
Glasgow 59, 60, 105
global cities 37, 38
global city thesis 74, 78
global networks and flows 65
globalisation 11,26; Anglo-American financial model 44; convergence of housing markets 88; diversification 64; economic, and cities 38; economic effects of 2, 1819, 20–1, 38; individual security 44–5; local government 154; meaning of 38; metropolitan 72; nation state 16–17, 19, 127; oligopolistic competition 39; opportunities for cities 19; social policy 125–6; uncertainty 44; urban concentration 20–1; urban development 112
Gloucester 105
governance, definition 26. *See also* urban governance
Greater London Council 94
Greece, progress to service society 113
Grenoble 59

Halbwachs, M. 76
Hamburg, as European metropolis 58; housing policy 134, 137–8; squatter movement 133
Hamnett, C. 79, 81, 87
Harloe, M. 11
Harvey, D. 4, 11
Helsinki 191, 192; desegregation 118–19; political groupings 123; unemployment 125; urban policy 28
hierarchy of European cities, 10–11
Hirshman, A. 29
homeless, protest movements 144–5
Hong Kong 2
housing: development 15; individual choice 90; low rental in Paris 86–7; housing market in European capitals 87; policy in Scandinavia 118–19; public policy 86; segregation 85–8; social segregation 86–7; transformation of inner cities 85–6. *See also* squatter movement

identity, sense of 98
Ile-de-France region, economic transformation of 80; and social segregation 80
immigration 13; integration in Paris and USA 89

Index

individual choice, and segregation 88–91
individualism 24
industrial development 15
industrial policy 20
industrial revolution, and European urbanisation 9
inequality: economic 13; incomes 79, 85; regional 37; social 13–14; welfare state 11617
infrastructure development 15
inner city areas, social transformation of 85–6
investment, transnational 38
Ipswich 106
Ireland, and business organisation 182; and progress to service society 113
Italy: business organisation 182–2, 184; elected mayors 28; industrial patterns 41; neolocalism 27; economic concentration 33; social structure of medium-sized cities 108–10; 'Third Italy' 101; urban policy 28; urban social movements 133

Jacobs, Jane 33
Japan, and globalisation 39

Kaijser, Arne 163
Konrad, G. 96

labour market 42–3, 4, 85
L'acteur et le systeme 154
Le Creusot 107
Le Figaro 168
Le Temps 168
Leeds 58, 105
Leipzig 59
Lenin 77
lex mercatoria 17
liberalism, economic, influence of 18–19
Lille 59, 60, 183
Lisbon 59, 60; mayors of 27
Liverpool 58, 60, 105
local government 25–9; administrative divisions 93; capacity of 153; changed conditions 153; collective action 154, 155; complexity of 93–4, 153; disenchantment with 154; elected mayors 27–8; impact of economic globalisation 154; leadership 94; origins 153; political ineffectiveness 92; urban services models 155–6. *See also* urban governance
local identity, in medium-sized cities 98
local taxation and self-segregation 90–1
localism, renaissance of 17

Lombardy 66
London 1, 2, 5, 6, 10, 14, 48, 105, 189–90, 196; economic concentration 33; fragmentation of urban governance 189; global city 11, 58; growth in air traffic 37; housing market 87; local government 94; social segregation in 76, 77, 78, 79, 81; weakness of urban governance 28
Lorrain, Dominique 192
Los Angeles 11, 14; and ethic enclaves 90
Lyonnaise des Eaux 192
Lyons 10, 183, 192; social divisions in 81

Machimura, T. 79
Madrid 60, 191; as European metropolis 59; housing market 87; social segregation 78
Maloutas, T. 79, 87
Manchester 5, 58, 59, 105, 186
Mantes 107
market, influence of 18–19
Marseilles 59, 60; social divisions in 81
Martin, Roger 167
Marx, K. 29, 77
Massey, Doreen 4
medieval city, characteristics of 7–8
Meijer, M. 59
metropolisation, effects of 2
metropolises 31; as major nodal points in global networks 91–2; concentration 55; confusion with cities 3; European 58–9; lack of identity 98; political weakness of 92; rarity in Europe 11–12; Simmel's analysis of 3
Mexico City 2
Michelin 166
middle class, expansion of 22; and medium sized cities 110–11; prominence of 24; and welfare state 115. *See also* class
Milan 5, 6, 11, 48, 60, 19; as European metropolis 59
Milton Keynes 106
mobility 19, 36; capital 42; increase in 23; labour 42, 47 n.5; and political involvement 93; population 12–13, 37
Mollenkopf, J. 11
Montpellier 59, 60
motor industry 47 n.14
Munich, as European metropolis 59

Nantes, concentration of foreigners 83; social divisions in 81

Naples 59, 192
nation state, and: decline in role of 74; development of 16; difficulties of 34; economic liberalism 18–19; effect on cities 8; influence of on European cities 12; national culture 17; problems in maintaining 16–17; restructuring of 178; weakening of 26, 127
neolocalism 27, 102
Netherlands, and urban gas supply 163–5; progress to service society 113
networks 21
new political economy 4, 29
New York 2, 14; ethnic enclaves 90; social segregation 77, 78, 79
Newcastle 60, 105
Norway, progress to service society 113; unemployment 120
Norwich 106, 192
Nuremberg 59

Oberti, M. 89, 91
Offe, Claus 181
out-of-town development 15

Palermo 60
Paris 2, 5, 6, 10, 14, 36–7, 189–90, 196; class and politics 94; economic concentration 33; fragmentation of urban governance 189; global city 11, 58; growth in air traffic 37; high concentration of foreigners 83; housing market 86; integration of immigrants 89; local government 94; low rental housing and social segregation 867; mayors of 27; political protest 94–5, 96; racial segregation in 82; social segregation 77, 78, 79, 80; weakness of urban governance 28
Park, R. E. 3, 76, 78
Perrier, J.-L. 96
Perugia 108, 109
Peterborough 106
Piedmont 66
Pittsburgh 157
Po valley 66
politics, and cities 91–5; disenchantment with 154; role of 156
Pont-à-Mousson 165, 166, 167, 168
Pont-de-Vaux 166
Portes, A. 89–90
Portugal, and business organisation 182
post-industrial phase 54–5
poverty, in cities 14

productivity 36, 42–3
Propex 166–7
protest movements. *See* urban social movements
public policy, and segregation 84–5
public transport 15
public utilities, as hallmark of European municipality 15. *See also* gas, water
public private partnerships 191
racial segregation 81–3. *See also* segregation

Radcliff–Maud Report 153
rail traffic, as measure of flows in urban network 65–6
railways, development of 162
Randstad 11; as European metropolis 58
Reading 106
regionalism 24; and economic policy 20; and neolocalism 27; and political organisation 25; renaissance of 17; significance of 34
regions, renewed political significance 74
Rennes 107, 192; social divisions in 81; and urban policy 28
Reynaud, Paul 168
Rhine–Ruhr zone 6
Rhineland, economic concentration 33
Ritter, Karl 50
road network, as measure of flows in urban network 66
Rome 11, 14, 192; as European metropolis 59
Rotterdam 189–90
Ruhr 11
rural developments 37

Saint-Gobain 165
Salais, R. 45
Salford gas scandal 171–4
Sassen, S. 11, 74, 82
Saunders, P. 77
Scandinavia: business organisation 183; homogeneity of cities 118; housing policy 118; local municipalities 27; political groupings 123–4; process of urbanisation 117
Scandinavian welfare state 117–26: city-based political action 122, 123; decentralisation 128; dominance in labour market 120; golden age of 117–19; impact of 123; inter-city competition 126; pressure for change 126; regional policy 126; response to global economic restructuring 119–

Index

20; role of cities 125; social policy 118 124–5, 129; support for consumers 120–2; unemployment 117–18, 120; women 121, 123
Schelling, T. 88
segregation in cities, political consequences 74–5; causes of 84–91; class 76–81, 88; communitarian impulse 88–90; consumer practice 88–91; education 90; housing 858, 118–19; individual choice 88–91; local taxation 90–1; public policy 84–5; racial 81–3, 88–9; social 2, 13–14; spatial 2, 13–14; working-class self-segregation 88–9
Seine–Saint-Denis department, social inequality 80
self-employment, rise in 40
service sector, and urban development 114–15
Seville 59
Shanghai 2
Shell Oil 163, 164
Simmel, G. 24, 31, 45; analysis of European cities 3
Singapore 33
single European market 66
Smith, Adam 33
social capital 24
social division, in cities 75–96; and city size 81; political consequences of 91–5; impact of welfare state 115; difficulties of analysis 81; international comparisons 78; within France 81
social exclusion 22, 144–5, 146, 148
social movements, urban. *See* urban social movements
social policy, role of cities 128–30
social stratification 77
social structure 21–4; of medium-sized cities 98–111; and retreat of industry 111; Britain 104–6; and cities of producers 101–2; France 106–7; and industrial cities 99–101; influence of middle class 110–11; Italy 108–10; and public-service dependent cities 103–4; sense of identity 98; and service cities 102–3; types of city 99
Sonobe, M. 79
South Africa, and racial segregation 83
South Korea, progress to service society 113
Spain, progress to service society 113

spatial images of urbanisation 48–73
squatter movement 133–4, 136–8
St Etienne 107
St Petersburg 5
stability, need for 71
Stalin, J. 77
state, role of 29
state intervention, and European cities 14–15; decline of 19
status stratification 22
Stevenage 106
Stockholm, as European metropolis 58; political groupings 123; unemployment 125
Storper, M. 45
Strasburg 192; social divisions in 81
stratification 22
Streeck, Wolfgang 181–2
students, and urban social movements 133–4
Stuttgart 10, 60, 192
suburban development 10
Swedberg, R. 29
Sweden, and: business organisation 183; diminishing role for local authorities 128; unemployment 118, 120; urbanisation of 117
Swindon 106, 192
Switzerland, progress to service society 113

Tabard, N. 80, 81
Tarr, Joel 157, 158, 160
technology, impact on cities 10; territorially based 42
territorial organisation 16–17
tertiary proletariat 79
The City 7–8
Therborn, G. 113
'Third Italy' 101
Thuillier, Guy 166
Tilly, Charles 16
Todd, E. 83
Tokyo 2
Toulouse 37; social divisions in 81
town planning, European tradition of 14–15
transnational trade 47 n.12
Trieste 60
Trigilia, Carlo 102
trust and economic efficiency 43
Turin 59, 60, 192

United States: black ghettos 89; characteristics of cities 8, 9; cities contrasted with European 8–16;

United States: (*cont.*)
economy, characteristics of 18;
homeless protesters 144;
home-owners' movement 140;
immigrant integration 89; progress to service society 113; property tax revolt 140; racial segregation 81, 84; welfare state 116–17
urban concentration 21, 41, 51–2
urban density 34–5
urban development, conditions for 65; growth of service sector 114–15; and welfare state 115–17; protest movements 139
urban disintegration, and mobility 23–4
urban ecology 3
urban governance 6, 25–9, 73, 91–6, 178–97; business interests and actors 180–8; changing forms of 179; European variations 28–9; forms of 26–7; incorporation of business 180; and integration of business 191–2; private actors 188–95; public-private partnerships 191; relations with state 195; strengthening of 190–1; urban interest groups 178–80; weakness of 189–90
urban identities, creation of 25
urban integration 6
urban networks 51–2, 65, 155; depiction of European 71; models of 69–70; and regional cohesion 69; spatial configurations of 66–70
urban regimes, theories of 192–3
urban renewal, and urban social movements 132, 134
urban research, trends in 1–2
urban services, development of models of 155–76; technical competition and stabilisation 157–61; development of institutional framework 161–5; developing principles and creating needs 165–71; legitimisation 171–4; role of crisis 174; political relevance 174–6; government functions 175–6; political intervention 175; power of firms 174; rehabilitation of politics 174–5
urban social movements: 1960s and 1970s 132–6; and left-wing radicalism 134; and recession 135; causes 132; composition 132–3; creation of urban counter-culture 134–5; methods used by 132; reaction to 134; squatting 133; fragmentation 135–6
urban social movements: 1980s 136–8; changing composition of 138; increasing local cooperation 136–7; squatter movement 136–7
urban social movements: 1990s 138–46; as alternative renewal agents 141–2; as responses to restructuring 139; city design 140–1; defence of home-values 140; formation of broad coalitions 140–1; fragmentation of 138–9; institutionalisation of 141–3; middle-class involvement 145; new poor people's movements 143–5, 148; prestige projects 140–1, 152 n.5; quality-of-life protests 139–40; urban development 140–1; urban revitalisation 141–3; 'Wagenburgen' 144–5, 152 n.10
urban social movements: and city-size 150; conflict produced by urban growth 146–7; disruption movements 148; fragmentation of 149–50; incorporation into political process 147, 149–50; as replacement of state provisions 148–9; response to labour market conditions 148–9; role 146–51; as strengtheners of civil society 151
urban sociology, foundations of 3–4; Marxist tradition of 4; and Max Weber 5
urban stability 8–11
urbanisation 52; extent of in Europe 11–12; historical origins 50; reversal of 52–3; spatial images of 48–73
utilities. *See* gas, water, urban services

Valencia 192
Veltz, P. 74
Venice 5, 192
Vernon products 155
Vienna 3, 5; as European metropolis 59

'Wagenburger' 144–5, 152 n.10
water, provision of in France 162, 165–71
Weber, Eugen 166
Weber, Max 2, 3, 15, 29, 47 n.2, 99; and cities 4, 6–7; uniqueness of European city 7–8
welfare state: crisis in European 129–30; different models 115–16; Scandinavian model 117–26; and shaping of cities 112–30; and urban social movements 132
welfare systems 22
Wilson, W.J. 77, 89
Winchester 105

Wolverhampton 105

youth protest movements 132. *See also* urban social movements
Yugoslavia, and ethnic segregation 84

Zentralen Orten 50
Zurich, as European metropolis 59; squatter movement 137